REPRODUCTION BY DESIGN

ODUCTION BY DESIGN

REPRODUCTION BY DESIGN

REPRODUCTION BY DESIGN

REPRODUCTION BY DESIGN

REPRODUCTION BY DESIGN

Sex,

Robots,

Trees, and

Test-Tube Babies

in Interwar Britain

Angus McLaren

University of Chicago Press | Chicago and London

Angus McLaren is the author of twelve books, including three previous University of Chicago Press titles: *Impotence: A Cultural History* (2007), *The Trials of Masculinity: Studies in the Policing of Sexual Boundaries, 1870–1930* (1997), and *A Prescription for Murder: The Victorian Serial Killings of Dr. Thomas Neill Cream* (1993). He is Distinguished Professor of History Emeritus at the University of Victoria.

The University of Chicago Press, Chicago 60637
The University of Chicago Press, Ltd., London

Printed in the United States of America

21 20 19 18 17 16 15 14 13 12   1 2 3 4 5

ISBN-13: 978-0-226-56069-4 (cloth)
ISBN-10: 0-226-56069-4 (cloth)

Library of Congress Cataloging-in-Publication Data
McLaren, Angus.
Reproduction by design : sex, robots, trees, and test-tube babies in interwar Britain / Angus McLaren.
p. cm.
Includes bibliographical references and index.
ISBN-13: 978-0-226-56069-4 (cloth: alk. paper)
ISBN-10: 0-226-56069-4 (cloth: alk. paper)
1. Technology in literature. 2. Reproductive technology—England. 3. Reproduction—Social aspects—England. 4. English literature—20th century—History and criticism. 5. Futurism (Literary movement) I. Title.
PR478.F87M37 2012
363.9—dc23
2011018998

∞ This paper meets the requirements of ANSI/NISO Z39.48-1992 (Permanence of Paper).

CONTENTS

Acknowledgments vii
Introduction 1

PART ONE Speculative Literature and Mechanistic Progress
1 "The standardized world we are facing and fearing": Sex and Futurist Fictions 9
2 "What is better, a car or a wife?": Automobiles and Modern Bodies 37
3 "A race of human machines": Robots and Reproduction 57

PART TWO Beyond the Predictive: Sex in Real Time
4 "A sort of animal or mechanic immortality": Glands and Eugenics 81
5 "A spinster and a syringe": Debating Test-Tube Babies 109

PART THREE Romantic Racialism
6 "Breeding a race apart from nature": Ruralists and Conservationists 137

Conclusion 157
Notes 165
Index 227

ACKNOWLEDGMENTS

In the interwar years, the eminent philosopher Bertrand Russell had a good deal to say about reproduction by design, but one line of his on another subject altogether has stuck in my memory. "Whether history is a science or not," he asserted, "it certainly can be an art, and I, for my part, value it quite as much for its intrinsic interest as for what it can establish in the ways of causal laws. I value it also for the knowledge it gives of human beings in circumstances very different from our own—not mainly scientific knowledge, but the sort of knowledge that a dog-lover has of his dog" ("Last Philosophical Testament: 1943–68," in *The Collected Papers of Bertrand Russell*, ed. John G. Slater and Peter Köllner [London: Routledge, 1997], 11:63). Some historians might bridle at such a statement, but I found it especially pertinent to my life over the past few years inasmuch as my research devoted to understanding the interwar debates over reproduction were interspersed with long refreshing walks with our poodle-wheaten cross. Whether or not my dog made me a better historian, Frida did provide me with welcome excuses to abandon my desk and escape to the woods.

Others assisted my endeavors in more practical ways. I owe an enormous debt of gratitude to the many friends and colleagues who assisted me in this undertaking. I first have to thank Brian Dippie for subjecting the manuscript—like so many earlier ones—to a careful and insightful reading. I only wish I had the space to respond to his many challenging ripostes. Countless students and colleagues provided me with encouragement and assistance. I particularly value the intellectual support offered by Judith Allen, David Anderson, Lucy Bland, Stephen Brooke, Hera Cook, Matt Cook, Catherine Ellis, Michael Finn, Carol Gerson, Christine Grandy, Lesley Hall, Kathy Mezei, Robert Nye, Eric Schaefer, Judith Walkowitz, and Patrick Wright.

Useful feedback from friendly audiences was received when parts of the study were delivered as lectures. Portions of this work were presented to the 2008 British Studies Association meeting in Cincinnati, Ohio, and as the 2009 Paterson Lecture to the Canadian Society of the History of Medicine conference in Ottawa, Canada, and as the 2009 Wellcome Lecture in the History of Medicine at Cambridge University. Chapter 4 in this study draws on chapter 8 in my earlier book, *Impotence: A Cultural History* (Chicago: University of Chicago Press, 2007).

I am very appreciative of the generous support of the Social Science and Humanities Research Council of Canada that enabled me to make the numerous overseas trips that this project required. My repeated stays in London were brightened by the hospitality and kindness of Susannah and Richard Taffler and Aimée and Michael Birnbaum. Special thanks goes to Michael Finn and Elizabeth Park for putting me up in Toronto. For their unflagging service, I owe a debt of gratitude to the staffs of the University of Victoria Library, the Woodward Medical Library at the University of British Columbia, the Cambridge University Library, the British Library, the Wellcome Institute for the History of Medicine, the Archives of King's College at the University of London, the University of Saskatchewan Special Collections, and the Robarts and Gerstein Libraries at the University of Toronto.

At the University of Victoria Andrew Rippen was a supportive dean, and Tom Saunders an understanding History Department chair. No one writes more engaging letters than Doug Mitchell; he and Tim McGovern made the production process at Chicago a pleasure. I am as ever indebted to Brian and Donna Dippie, whose boundless generosity made my continual shuttling between Victoria and Vancouver possible. Although now an emergency room physician in Toronto, Jesse still keeps me from taking myself or my projects too seriously. Arlene remains my main critic and supporter.

INTRODUCTION

Though I am tempted to claim that the following study was carefully crafted as a response to a number of recent books linking sexuality and modernization, I have to admit it was first sparked by my stumbling across a few mysterious references in the archives of King's College, London.[1] Having written several books on the history of contraception, I decided that it might be worthwhile to go through the private papers of one of the more extreme British eugenicists of the interwar period, Reginald Ruggles Gates, the first husband of the birth controller Marie Stopes.[2] To my puzzlement, I found in his files some letters and circulars from an organization called the "Men of the Trees." The name conjured up Tarzan-type figures, but it was, I learned, an association dedicated to the planting and protection of trees. The surprising thing was that Gates was a supporter and its treasurer was Ursula Grant Duff, one of the English Eugenics Society's chief organizers. Why, I wondered, were eugenicists like Duff and Gates involved in the apparently benign Men of the Trees movement? The answer seemed to be that both eugenicists and conservationists shared a concern for healthy reproduction—be it of trees or people. In a mechanizing, modernizing world, they found it increasingly difficult to sustain a Darwinian faith that "nature" could be relied upon to promote healthy fertility; what was needed was reproduction by design.

It then occurred to me that a number of other questions that had attracted my attention—including the emergence in the interwar years of hormonal therapies and artificial insemination—had elicited similar debates. In noting how some of these discussions overlapped, I came to realize that what at first glance appeared to be quite disparate issues were best understood if viewed as different facets of the same elite cultural preoccupation.

The central argument of this study is that reproduction was a key site for many of those debating the merits of the modern mechanized world. To understand why, the book draws on novels, plays, and science fiction, and the work of biologists, psychiatrists, sexologists, and medical scientists. It highlights a cluster of ideas and images, fears and fantasies revolving around the notion that the boundaries between natural and artificial in reproduction were collapsing—that ecological and scientific advances were bringing about a crisis in the relationship of humans, machines, and the environment.[3] Taking such an approach raises, of course, the issue of proportionality. Setting out to find those who discussed the linkage of modernity and reproduction, and discovering them in large numbers, one runs the risk of imagining that the subject erupted everywhere and was on everyone's mind. Obviously this was not the case. This book does not argue that the works it examines were in some way "representative" of interwar writing, but it does assert that they represented an important and as yet unexamined strain in British culture.[4]

Though the study is centered on the 1920s and 1930s, it necessarily ranges at times back to the nineteenth century and forward to the 1950s. It argues that interwar discussions of sexuality, reproduction, endocrinology, eugenics, and environmentalism were hopelessly entangled. Such an approach posits that novelists, eugenicists, birth controllers, demographers, doctors, conservationists, or indeed anyone talking about the future of sex and reproduction, were in effect producing "science fictions." That is to say, many scientific theories and notions of the future, especially the forebodings, were in fact crystallizations of current social concerns.[5]

Why did such debates occur? In the nineteenth century, doctors could not safely penetrate the body. In the twentieth century, they were equipped with technologies that, in addition to allowing them to repair the damaged, induced them to attempt to render "normal" the deviant and assist in the creation of life. Impressed by a number of remarkable social changes—in particular advances in medical science, the mass employment of birth control, and the emancipation of women—a variety of commentators in the early twentieth century came to espouse the notion that sex and gender were not timeless categories; they could be manipulated. They had a past and accordingly a future. The reproductive body therefore became, as it never had been before, a central site for political and cultural contention. Male scientists reveled in the possibilities of "creating life"; feminists worried that women's bodies would be increasingly investigated and controlled. Despite vociferous opposition, research was undertaken that ultimately resulted in a range of dramatic technologies including chemical contraception, artificial insemination, and hormonal treatments being naturalized and legitimized.[6]

Both a cause and an effect of such experimentation was the vigorous debate of much of what we take to be the hallmarks of our twenty-first century, liberal culture—homosexuality, recreational sex, contraception, abortion, euthanasia, hormone replacement therapies, sex change operations, and in vitro fertilization. H. G. Wells, Bertrand and Dora Russell, J. B. S. and Charlotte Haldane, C. E. M. Joad, Aldous and Julian Huxley, Anthony Ludovici, Naomi Mitchison, Vera Brittain, Marie Stopes, C. P. Blacker, J. D. Bernal, Enid Charles, and countless novelists and science-fiction writers presented the sexual body as a particularly privileged carrier of meanings. They were especially preoccupied by the question of how medical science would affect the family and reproduction, and their hopes and fears informed popular interwar discussions of everything from birth control and test-tube babies to euthanasia and environmentalism. The apparent collapsing of boundaries separating the natural and the artificial in reproduction was both frightening and inspiring. Many sought to shore up the naturalness of gender divisions and maternity, while making reproduction an object of manipulation. Those in Britain who tackled such issues did not simply ask how such changes were currently affecting sex and reproduction, they were as preoccupied in pondering what their future consequences might be.

It has been famously said of the moderns: "They are moved at once by a will to change—to transform themselves and their world—and by a terror of disorientation and disintegration, of life falling apart. They all know the thrill and the dread of a world in which 'all that is solid melts into air.' To be modern is to live a life of paradox and contradiction."[7] Such a paradox was evident in many British works that discussed intervening to create the "normal" or the "natural." Progressive commentators commonly spoke of the need for the body to be rationally reconfigured. They frequently drew their metaphors for reproduction from industry. Social and political ideas thus shaped the reproductive process.

Eugenicists set much of this agenda in raising the specter of civilization itself producing threatening armies of degenerates. We know a great deal about eugenics in Britain. Biographical accounts of its chief propagandists, chronicles of the Eugenics Society, and accounts of attempts to pass eugenically informed legislation have all appeared.[8] The homeland of eugenics, ironically enough, proved itself to be among the most resistant of European nations when it came to passing such laws. Should it be concluded that in Britain eugenics was a failed movement? This study argues that if one sets aside the issue of legislative accomplishments and instead traces the success with which eugenic notions percolated through the culture, a more nuanced account emerges. The pervasiveness of eugenic thinking was perhaps best revealed by the fact that interwar writers of so many different stripes took on board the suggestion that the race could be improved. To appreciate Marie Stopes opening her birth control clinic in 1921, or Karel Čapek putting on *R.U.R.*, the play that introduced the word "robot" to the English-speaking world in 1923, or J. B. S. Haldane explaining in a tract of the same year why test-tube babies would soon be possible, or Norman Haire describing in 1925 how the transplantation of testicles and ovaries could counter senility and same-sex attractions, required some understanding of eugenics.[9]

Eugenicists succeeded in assuring that almost every discussion of the future included speculations about sexual ideas and practices. Most commentators ultimately judged the moral standing of their future imagined world, not so much on its views on race or class, as on its attitudes toward the family and reproduction. Would women produce fewer children? Would babies be artificially conceived? Would men become sexually rejuvenated? Would the form of the family change? Would the race be improved? Would homosexuals win acceptance or be "cured" by hormonal treatments? Would an urban mechanized world obliterate nature? Many were intensely ambivalent about, some resolutely hostile to, the forces of modernity. Optimists embraced most technological and scientific advances yet worried that altering existing class, gender, and racial relationships might lead to changes for the worse, to degeneration, and that at the very least an "Americanized" culture might result. Conservative commentators were openly suspicious of the emerging mass society, of commodification and of Fordism, and sought ways to counter it. All advanced their own designs for living.

In *Modernity and the Holocaust* (1989), Zygmunt Bauman argues that the Nazis, far from being opposed to all aspects of modernity, embraced the modernists' belief in the need to plan and perfect nature. Gardening and medicine served as the archetypes of this engineering attitude, which held that human life should be subjected to planning and administration.[10] The fact that British writers were also vigorously debating such assertions has been slighted. A large scholarly literature exists on the sex radicalism of Weimar Germany and the early Soviet Union while much of the lively British discussion has been overlooked. Many historians have stressed the tenacity of antimodern forces in Britain as opposed to Germany, but it depends on what aspect of society one looks at. The argument can be made that nowhere in the interwar world was there a more modern discussion of sex and reproduction than in Britain. Though Britain was generally viewed as having a conservative sexual culture and a repressive penal code, authors who speculated on how gender relations and family life might change in the future were permitted a good deal of freedom.[11]

Who were discussing such topics? Much has been written about the bright young things of the 1920s. This study will not rehash the well-known affairs of the rich and famous. The argument rather is that what came to be understood as the idea of "modern sexuality" was not so much the brainchild of sensualists as the product of a line of often eugenically inspired rationalists who, out of a concern for the future of Britain's population, sought the disciplining and regulation of the body. For many who debated the merits of modernity, eugenic notions provided an overarching answer. Both the progressives who defended birth control and the conservatives who lauded the importance of pure blood drew on eugenic arguments. In Britain, as elsewhere in the Western world, modernity was commonly represented by the futurist notion (heavily colored by eugenics) of streamlining humans—of making people more like machines and machines more like people. While American science-fiction writers were obsessed with extraterrestrials, rocket ships, and death rays, the British were hypnotized by the possibilities and pitfalls of harnessing biological change.

Huxley's *Brave New World* (1932) remains the best-known depiction of a scientifically planned and controlled society. It is usually portrayed as being predictive of future trends rather than reflective of current cultural preoccupations. Huxley would disagree. "Prophecy is mainly interesting for the light it throws on the age in which it is uttered," he insisted. "Our notions of the future have something of that significance which Freud attributes to our dreams. And not our notions of the future only: our notions of the past as well."[12] This study will accordingly begin in part 1, "Speculative Literature and Mechanistic Progress," by tracing fictional investigations of the impact of modernization on sexuality. Chapter 1 demonstrates how much of futurist literature focused on the topics of femininity, masculinity, birth control, and reproduction. Writing about the future was, of course, a classic means—that can be traced as far back as Campanella—of legitimating an analysis of sex relations.

The subsequent chapters turn to key debates over the modernization of sexuality in the first half of the twentieth century. How was the human/machine relationship portrayed? Chapter 2 explains why probably the most ubiquitous representation of the impact of machine culture was the sexualization of the motorcar on the one hand and the automobilization of humans on the other. No other machine was credited with having such an impact on sexual relations. Pessimists foresaw the increased production of motorcars resulting in the decreased reproduction of people. Given such fixations on the blurring of the machine/human boundary, it followed that the arrival of the robot raised even greater worries. The extent to which such concerns about the threat of mechanical men were transparently eugenic in nature is discussed in chapter 3.

In part 2, "Beyond the Predictive: Sex in Real Time," popular literature is again exploited as evidence of the secularization of science, but the focus is on the works of medical scientists. Chapter 4 shifts the attention from robots to real men, showing how hormones were first envisaged as being able to improve male sexual performance and then extended to the notion of generally promoting social efficiency. This would include their employment to enforce or restore sexual "normality." Patients like automobiles could, it was claimed, be repaired and rejuvenated by skilled experts. It might be hard to criticize therapies that promised to make men more manly and women more womanly, but how did the interwar public view "test-tube babies" resulting from artificial insemination? Was love displaced by mechanics when husbands were replaced by syringes? In chapter 5, an analysis is made of the debate over whether such interventions should be condemned as unnatural or lauded as a means of simply assisting nature.

Finally, the inevitable backlash against some key aspects of modernization is dealt with in part 3, "Romantic Racialism." There already existed a deeply ingrained antimodernist strain in British culture, but with the onslaught of mechanization in the interwar years, "nature" itself won a new generation of defenders who linked the healthy reproduction of trees and the healthy reproduction of humans. They presented themselves as courageously opposing the advent of the "brave new world" of urban consumerism. Chapter 6 seeks to explain why those who claimed that the coming of mass society entailed ecological disaster could not avoid linking up sexual, eugenic, and conservationist concerns.

A large and eccentric cast of characters including seductive motorcars, rebellious robots, friendly trees, and timorous test-tube babies populates this brief study. Its goal is to better understand why in a remarkably short space of time modernizers (of a variety of stripes) succeeded in advancing the argument that the protean forces of sex and reproduction had to be subjected to planning and control. They, of course, did not win the debate—it is still going on. They did launch it, however, and we have to know something of its history if we present ourselves as qualified to participate in the current round of discussions.

PART ONE

Speculative Literature and Mechanistic Progress

CHAPTER 1

"The standardized world we are facing and fearing"

Sex and Futurist Fictions

In the first pages of *Brave New World* (1932), Aldous Huxley's famous portrayal of a future dystopia, the author plunges us into a world where the state controls every aspect of human reproduction. In minute detail, he describes the Central London Hatchery and Conditioning Centre, with its Fertilizing Room, Bottling Room, Social Predestination Room, and Decanting Room. Most women in his envisaged society are induced to have their ovaries removed. Their eggs are harvested, inspected for abnormalities, and mixed with spermatozoa. Once conception occurs, the fertilized egg is incubated, fed, bathed in hormones, its sex determined, and the crucial decisions made—will it be twinned? Will it have its development arrested? And, if female, will it be made into a freemartin? Finally, instead of being born, babies are "decanted." In this society, carrying a child to term is taboo, its citizens conditioned to be disgusted by the idea that reproduction was once "viviparous." Now, Huxley has the director of the hatchery hailing ectogenesis—that is the conception and gestation of babies outside the womb—as "The principle of mass production at last applied to biology."[1]

A prolific writer, Aldous Huxley (1894–1963) was hailed as the voice of postwar upper-middle-class youth. A string of comic novels about cynical young men first established his reputation as an attacker of taboos. He was to move ultimately on from dandyism to spiritualism, but it was *Brave New World* that was to win him lasting fame. As a member of an intellectual dynasty (his grandfather Thomas Huxley was Darwin's most famous defender), he was, unlike most mandarins, not hostile to science. Nevertheless his doubts about its ability to guarantee social progress led some like H. G. Wells to regard him as a pessimist.[2]

Innumerable studies have been made of Huxley's classic, with particular attention paid to his success in predicting future social and scientific developments.[3] What has been slighted have been the ways in which he drew on contemporary concerns about reproduction. *Brave New World* was far from being original. It was only the most extreme and ultimately the best-known fictional manifestation of a cultural malaise, a British ambivalence about the ultimate benefits of the encroachment of science on human life. Even when sensationally raising the specter of Taylorism (or scientific management) applied to biology—suggesting the coming of a world in which natural childbirth would be regarded with horror and replaced by oophorectomies and ectogenesis—Huxley was drawing on concerns voiced by many others. To understand why they were broaching such issues, this chapter will sketch out the sexual challenges they believed that their modernizing society was producing—including restless women, incapacitated men, and declining fertility. The goal is to explain why so many believed that changes in reproduction and gender relationships posed serious dangers, and why the intervention of eugenics and medical science was regarded both as a cause and cure of such problems.

Between the 1870s and the 1940s Britain witnessed a surge in such futurist writing. A small army of writers tackled the question of what impact modernization might ultimately have on sex and the family. Those who addressed these issues ranged from highbrow novelists to middlebrow writers of science fiction. The late nineteenth-century speculative fictions generally reflected a belief in progress and technology.[4] In the twentieth century, a shift occurred, with a decline in the utopian interest in social or political planning, and a growing fixation on the possibility of the biological penetration and transformation of minds and bodies. What if the ambition was not just to produce planned communities but to plan and control nature itself? Building motorcars on assembly lines or scientifically organizing the planting of trees was all well and good. It was the question of whether births should be planned and controlled that sparked the fiercest responses. Debates about the merits of subjecting humans to scientific surveillance and intervention necessarily and inevitably focused on reproduction.[5]

Those offering predictions about how sex and reproduction might change offered a variety of scenarios. Early futurist literature optimistically tended to assume that in an egalitarian future world women's reproductive decisions would not be constrained by social concerns. In the American Edward Bellamy's *Looking Backward: 2000–1887* (1888), a cooperative society of the year 2000 continues gender norms, but women have no "unnatural rivalry" with men. If they are "allowed" by men to work—since it keeps them healthy—sexual differences are still respected. Employment does not deter women from marriage—indeed one has to be a wife and mother to obtain the highest status.[6] Bellamy argued that if sexual selection were to operate freely, it was necessary that women not marry for money. Nevertheless he credited his economically independent women with continuing to be charmingly feminine. He asserts that coquetry is ended, yet girls continue to blush. If women are purportedly "free," before marrying they still seek their fathers' consent.

The English socialist William Morris responded to Bellamy with *News from Nowhere* (1890). His too is a utopia where domestic tyranny, sexual ignorance, and "commercial marriages" are unknown. Women are happy to be wives and mothers. The idea of the "superior" women to emancipate their sex from the bearing of children is now recognized as folly. As maternity in a cooperative community poses no hardships, the free woman "has far more instinct for maternity than the poor drudge and mother of drudges of past days."[7]

In *A Modern Utopia* (1905), H. G. Wells likewise imagined sexual desires being harnessed for the benefit of the community. At the moment, love was made "too elaborately," too much erotic brooding occurred. Overindulgence was followed by demoralization and excesses; promiscuity led to social instability. Such dangers were skirted in utopia. "And, in the matter of love, a straight and clean desire for a clean and straight fellow-creature was our Founders' ideal. They enjoined marriage between equals as the samurai's duty to the race, and they framed directions of the precisest sort to prevent that uxorious inseparableness, that connubiality which will reduce a couple of people to something jointly less than either."[8] The novelist and historian H. G. Wells (1866–1946) was commonly regarded as the father of British science fiction. He was the best-known writer of the generation who prior to World War One brought to a close a form of reassuring futurist writing that looked forward to an idealized state. His fame as a prophetic thinker peaked about 1910, but he continued to churn out didactic accounts of the social and political functioning of future societies that demonstrated his optimistic faith in science.[9]

Responding to the rise of organized labor and the creation of socialist parties, twentieth-century writers made more pessimistic predictions. Drawing on reports of advances in the biological sciences, the New Zealander Godfrey Sweven

(John Macmillan Brown) in *Limanora: The Island of Progress* (1903) describes a society where reproduction is supervised by the state. Only the best are allowed to propagate while the diseased are sterilized. Childbearing is carried out to fill vacancies. The character of the embryos are known and their development guided before their birth.[10] In attacking the idea of such a future socialist dystopia, the anonymously penned *Backwards and Forwards* (1905) portrays a totalitarian state with its state schools and post offices, its legislation on haircuts and facial expressions. It even has written rules of courtship.[11] Women, the author asserts, become slovenly drabs since their love of approbation has vanished. They had once been the more vain sex, but since marriages are now arranged they no longer compete for attention. Indeed the state sorts out couples to provide biological uniformity, gives them numbers instead of names, and has them live in barracks.[12]

One key question haunted such speculative writing—would women change? In the earlier utopian or futurist accounts, women are presented as essentially passive. Countering such hopeful depictions of a future in which sexual harmony would reign was a long line of works inspired by the male fear of women ultimately rebelling and seizing power, a prospect inspired by the current suffragist campaign. Walter Besant's *Revolt of Man* (1882) begins with women in control, but men rise up against the female dictatorship and even women recognize their mistake of seeking equality.[13] Billing itself as "Appointed for use in the National Schools of Japan, Tokio, 2005," Elliot Evans Mills's *The Decline and Fall of the British Empire* (1905) attributed the domination of the country by the town as responsible for Britain's twenty-first-century enfeebled health, undisciplined hooligans, excessive taxes, and emancipated women.[14] Similar antifeminist accounts include Allan Reeth, *Legions of the Dawn* (1908); Jesse Wilson, *When the Women Reign: 1930* (1909); A. C. Fox-Davies, *The Sex Triumphant* (1909); and Anon., *When Woman Rules! A Tale of the First Women's Government* (1923).[15] Comic versions of the threat to gender norms also appeared. In *John of Jingalo: The Story of a Monarch in Difficulties* (1912), Laurence Housman, presented a king supporting women's suffrage after falling on his head.[16] The Canadian humorist Stephen Leacock envisaged a world in which men read the fashion columns and took hours to dress for dinner while women threw on their clothes and demanded to be paid for their domestic services.[17] Arguments made in favor of the clear distinction of the sexes were pushed to ludicrous lengths. In the dreary *An Unknown Land* (1942), H. L. Samuel imagined a search for Bacon's New Atlantis resulting in the discovery of Bensalem, a cooperative society in which suturization is employed to increase the brain size of newborns.[18] Those who oppose such operations on moral grounds form an inferior and separate caste. It goes without saying that men's skulls are always larger than women's.

Other writers produced far more radical accounts of possible shifts in gender relations.[19] One popular genre was the positive depiction of women finally coming to political office.[20] Sexual issues were usually skirted in such novels, but in Lady Florence Dixie's *Gloriana; or, The Revolution of 1900* (1890), a cross-dressing young woman becomes prime minister with the result that women's emancipation is achieved, the population problem is solved, and the blight of abortions ended.[21] A 1930s version of such a tale was Elise Kay Gresswell's *When Yvonne Was Dictator* (1935). It told of a young woman whose charisma wins her the love of the people who make her prime minister and finally dictator. The economic situation is initially so dire that the unemployed engage in mass suicide. Yvonne finds the answer in Malthusian economics. In her maiden speech at the age of eighteen, she says she would make it a crime for a man to father more children than he can provide for. She insists relief be given in kind. Work, not the dole, is the answer. Once dictator, her platform consists of the revision of marriage contracts, a women's charter, extension of divorce, establishment of labor colonies for the unemployed, compulsory retirement, the disestablishment of the church, national health insurance, and equal pay and working conditions. Thanks to a motherhood strike, a fall in births occurs, heralding a revival of the economy.[22]

Much was written openly about the prospect of women entering politics. More care was taken by writers in broaching the issue of releasing female desire, but futurist fears of sexually liberated women had a long history. In Edward George Bulwer-Lytton's *The Coming Race* (1871), an adventurer who discovers an underground society is most disturbed by the power women enjoy there. Because they are socially superior, their faces are "devoid of the softness and timidity of expression which give charm to the face of woman as seen on earth above."[23] The narrator is stirred by one girl's visage "because it looked less bold—less conscious of female rights."[24] In this world, women wear red if they prefer to be single, and gray when in pursuit of men. While males play the coquette, females are sexually aggressive and make "immodest" overtures. The erotic interest taken in the narrator by Zee—"my host's highly informed and powerfully proportioned daughter"—is dangerous as she could easily reduce him to a cinder.[25] Women have full equality here, even the power to destroy their spouse; fortunately, the one time a woman destroyed her spouse, women were so horrified that they swore to never do it again. Similarly divorce and polygamy, though allowed, are not used.

As one approached the twentieth century, the looser female morals that writers had attributed to other times and places were discovered closer to home.

H. G. Wells in his scientific horror story *The Island of Dr. Moreau* (1896) reported that when half-human beasts reverted to their animal state, the females were the first to abandon monogamy and return to promiscuity. His narrator escapes the island, yet once back in civilization feels people are much like animals. Prowling women "mew" after him, and the blank-faced commuters seem hardly human.[26] For Wells, the civilized woman was a "sexual specialist," more sexualized than her primitive sisters. "Arrayed in what she calls distinctly 'dress,' scented, adorned, displayed, she achieves by artifice a sexual differentiation profounder than that of any other vertebrated animal."[27] Too often, he declared, she was an "unwholesome stimulant" on man.[28]

A common trope in futurist writing was to imagine female sexual mores being undermined as a result of a disastrous war. In reporting on the surges of blights such as venereal disease and divorce following the Great War, the newspaper press popularized such notions. Edward Shanks's *The People of the Ruins* (1920) portrayed the London of 2074, a tragic wreck of a society that had suffered decades of warfare.[29] The narrator has a romantic tie with a young woman, but Shanks does not provide a full account of sex or gender changes as did Cicely Hamilton in *Lest Ye Die* (1922). Though starting with a similar scenario of a war-ravaged world reverting back to savagery, she differed in highlighting the emergence of sexually aggressive women. One flings herself at the hero—Theodore Savage—who has just enlisted. He's embarrassed; she's excited by the conflict. With the collapse of society, some women prostitute themselves. Savage picks up a girl and they enter into the "married state," though the author notes the absence of legal sanction. On her death he takes another wife.[30] Such themes were echoed by the Scottish Marxist Lewis Grassic Gibbon in *Gay Hunter* (1934), which provided an account of England after a devastating atomic war. There is again a reversion back to a hunter-gatherer world where the naked narrator flees vicious fascists and shamelessly mates with a female friend.[31]

Surprisingly enough, Huxley, who questioned so many pieties, presented gender relations in his dystopia as unchanging. He was clearly not especially sympathetic to women.[32] All the scientists in his *Brave New World* are male; the chief female character is the stereotypical dumb blonde.[33] Lenina, described as "wonderfully pneumatic," first appears pulling down the zippers on her jacket, trousers, and panties. Zippers, in the 1930s considered modern and masculine, make the sexually aggressive woman's body immediately accessible. If such a portrayal was morally subversive, it remains the case that in Huxley's account all that women gain from centuries of scientific progress is that men unapologetically viewed them as meat.[34]

There were feminist futurists who boldly tackled the sex question. Responding to misogynists, Dora Russell (wife of the philosopher Bertrand Russell) presented in *Hypatia* (1925) a spirited defense of the sexually liberated women. Women of the 1920s were, she asserted, healthier and freer than ever before. The war led to their sexual emancipation, and they now were willing to admit their enjoyment. Russell went furthest in arguing that sex provided the woman with a real understanding of others and did not need to be justified by marriage or children.[35] She even argued in favor of polyandry.[36] Discussing the future of morals, the philosopher C. E. M. Joad agreed that economic change was affecting marriage. Women no longer had to sell themselves, and the line between married and unmarried was increasingly blurred. Birth control was spreading, and along with it came an end to the sexual double standard. If women were revealed to be no more monogamous than men, that did not mean that orgies ensued. Opposition to such changes, Joad argued, were based on fear of pleasure, not any actual evidence of danger to the race. He held up the example of the Soviet Union where, as a result of morals being more relaxed, sex was not as overvalued as it was in the West.[37]

Where did men figure in such speculations about the future? In *Brave New World,* elite men are preoccupied by sex and leisure time activities while the working masses prefer to consume drugs rather than fight to win their "manhood."[38] The state enjoins men to be like machines: "There must be men to tend them, men as steady as the wheels upon their axles, sane men, obedient men, stable in contentment."[39] The fear that the progress of civilization might empower women but weaken men was a common trope in futurist literature. H. G. Wells typically worried about gender boundaries being blurred. In *The Time Machine* (1895), the narrator's first concern on entering a future society is, "What if in this interval the race had lost its manliness, and developed into something inhuman, unsympathetic, and overwhelmingly powerful?"[40] In fact, he meets frail creatures, the Eloi—hairless, girlish vegetarians. They are so small and beautiful he finds it difficult to tell the sexes apart. Wells, himself a puny little man, represents the evil Morlocks as no more manly. His narrator is revolted by these "human rats": "soft creatures" with lank fingers, clammy little hands, and whispering voices. Gender ambiguity was similarly a focus of Edwin Lester Arnold's *Lieut. Gulliver Jones: His Vacation* (1905). His narrator discovers Mars to be inhabited by a caste of genderless toilers. In this world, they draw lots to select their marriage partners. In propagating, they simply do their duty and have no concept of father or mother. He asks the pretty boys their sex; they want to know why it's so important to be one or the other. Fortunately the Martian to whom he is most attracted turns out to be female.[41]

A number of writers took homosexuality as the logical manifestation of masculine decline. Max Nordau's prognosis was that in the twentieth century degeneration and hysteria would spread. Sexual psychopathy is soon so advanced that men wear clothes that in color and cut resemble feminine attire, while women in turn sport masculine dress. Modesty and restraint vanish. The complete perversion of social norms is finally obtained: "The demand of persons with the 'contrary' sexual sentiment that persons of the same sex can conclude a legal marriage has obtained satisfaction, seeing they have been numerous enough to elect a majority of deputies having the same tendency."[42] H. G. Wells, in *The Shape of Things to Come* (1933), has a future historian disdainfully comment on how chaotic twentieth-century society was as compared to that of the constrained Victorians. "As the gravity of economic and political problems increased and the structural unsoundness of the world became more manifest, sexual preoccupations seem to have afforded a sort of refuge from the mental strain demanded by the struggle. People distracted themselves from the immense demands of the situation by making a great noise about the intensifications and aberrations of the personal life. There was a real propaganda of drugs and homosexuality among the clever young."[43] In *Man's World* (1926), Charlotte Haldane (wife of the geneticist J. B. S. Haldane) describes a character who "tended to be intermediate sexually." His "submasculinity"—blamed on his mother not doing the right antenatal exercises—results in his mystical impulses and sterility. For Haldane, it is essential that men and women be normal if they are to create "a self-conscious race."[44] In fact, interwar science fiction infrequently portrayed homosexuality, and Olaf Stapledon's *Odd John* (1935) was especially unusual in nonjudgmentally having its chief character pass through a homosexual phase.[45]

Futurist literature repeatedly harped on the question as to what purpose sex would serve. Strikingly enough, when *Brave New World* appeared, the American magazine *Amazing Stories* panned the book for saying too much about sex and being too negative about science. "From the point of view of the science-fiction fan, this book is a decided flop. Its contents, which at times almost become obscene, but are at all times supercharged with sex, will undoubtedly bar it from circulation in Boston, but the book may serve to call the attention of a great many readers to the fact that there is a class of fiction in existence which deals with scientific subjects."[46] One can see why the book was regarded by some as indecent. Huxley presents a society where monogamy and romance are condemned, while sex education and erotic play are obligatory. Young women are embarrassed to admit that they are not promiscuous. Men chew sex-hormone gum and swap partners. Drug-fueled orgies ("orgy-porgy") replace religion. But did Huxley favor such behavior? The biochemist and sinologist Joseph Needham pointed out that in the novel sex was used by the state as a diversion and a means of control. Social stability was achieved by conditioning and "by allowing 'unlimited copulation' (sterile, of course) and unlimited sexual gratification of every kind."[47] The philosopher Bertrand Russell preceded Huxley in foreseeing such a scientific world in which ardent affections among the elite would be suspect whereas the lower orders would be provided with frivolous entertainments. "Love-making among the sterilized will be subjected to no restrictions either of law or of public opinion, but it will be casual and temporary, involving none of the deeper feelings and no serious affection."[48]

H. G. Wells was perhaps the most vocal in predicting that in the future Victorian sexual restraints would ease. In *Anticipations* (1902), he claimed, "So few people seem to be leading happy and healthy sexual lives that to mention the very word 'sexual' is to set them stirring, to brighten the eye, lower the voice, and blanch or flush the cheek with a flavour of guilt. We are all, as it were, keeping our secrets and hiding our shames." The state, he insisted, only had a right to interest itself in procreation; recreational sex should be free. The sexual question he compared to one's liking for golf. "In each case it would be for the medical man and the psychologist to decide how far the thing was wholesome and permissible, and how far it was an aggressive bad habit and an absorbing waste of time and energy. An able-bodied man continually addicted to love-making that had no result in offspring would be just as silly and morally objectionable as an able-bodied man who devoted his chief energies to hitting little balls over golf-links."[49]

So too in Godfrey Sweven's *Limanora: The Island of Progress* (1903) sex ceases to have its allure. "The lustful had been exiled and it was easy to eradicate from the natures of those that remained all trace of sexual passion, and with it all pruriency. The chief purpose of sex in nature, that of propagation of the family, became its sole purpose."[50] Since sex feelings are gone, so too is prudery; citizens dress in diaphanous gowns that conceal nothing. Their scientists conclude that sex was an accident, which not all species needed. It had once played its role in evolution, but now it has to be controlled if the race is to progress. John Gloag in *To-Morrow's Yesterday* (1932) went so far as to imagine the emergence after millions of years of catlike beings who, unlike lascivious humans, are not distracted by sex except during set seasons. "As our kind developed their brains they conquered fear and lust, for the first scientists soon learned to mechanize the reproductive process, so that in half a million years sex was altogether silenced, leaving us a freedom men had never known."[51]

James Hilton's *Lost Horizon* (1933), the best seller responsible for introducing the term "Shangri-La" to the English vocabulary, was yet another example of the British fascination with a faraway society protected, not only from the contamination of "dance-bands, cinemas, sky-signs," but also from the "hard, mocking, sex-thirsty voices of women" of the outside world.[52] The inhabitants of this Himalayan retreat, seeking tranquility through yoga, purportedly recognize no sex or race distinctions, though its founder and lama is a French priest and his successor is to be an Englishman. In Shangri-La, where the monks conveniently speak English, Father Perrault states that the Chinese and Tibetans are rather insensitive so do not live beyond one hundred. "Our best subjects, undoubtedly, are the Nordic and Latin races of Europe; perhaps the Americans would be equally adaptable."[53] The narrator finds the female lama has virtues lacking in the Western woman and loves his little Manchu. Unlike the "passion-driven races" she knows how to "calm the throb of desire" and pursue serenity.[54]

In the twentieth century, any discussion of sexuality had to take into account the availability of birth control. In 1918, Marie Stopes published *Wise Parenthood,* which directly broached the issue of birth control by providing diagrams of the reproductive organs and descriptions of a variety of contraceptives. To make such devices accessible to the working class, in March 1921 Stopes opened her Mother's Clinic in Holloway Road, London. The chief aim of Stopes's Society for Constructive Birth Control and Racial Progress, also established in 1921, was to pressure government officials into taking over the responsibility for such work.[55]

Society had once relied on women's fear of pregnancy as a natural method of policing sexual behavior. Futurists were divided on the possible impact of easy contraception. Optimists predicted the coming of healthier and happier families. Pessimists viewed the spread of birth control information as potentially as threatening to morality as the ravages of war. The debate over the pros and cons of fertility control had, of course, predated modern contraception. H. G. Wells's *The Time Machine* includes a clear enunciation of the Malthusian population theory: "Where population is balanced and abundant, much child-bearing becomes an evil rather than a blessing to the State; where violence comes but rarely and offspring are secure, there is less necessity—indeed there is no necessity—for an efficient family, and the specialization of the sexes with reference to their children's needs disappears."[56] The problem, as presented by Wells, was that with too much civilization the weak survive. "Strength is the outcome of need; security sets a premium on feebleness."[57] Spared threats and hardships a slothful society emerges that is, "against connubial jealousy, against fierce maternity, against passions of all sorts." The Elois now do nothing save entertain themselves and make love "in a half-playful fashion."[58] In short, the population is checked too well. Wells returned to the subject of population control in *The Dream* (1924). From two thousand years in the future, beings look back to the 1900s and marvel at the society's gloomy religion, poor education, and swarms of "unpremeditated children." "Love was a disgrace, a leering fraud, a smutty joke."[59] Divorces, abortions, and deaths in childbearing are all attributed by Wells to attempting to adhere to an impossible moral code.

In Wells's *Men Like Gods* (1923), birth control works because it is employed eugenically. Mr. Barnstaple drives his little yellow car into the future. He finds a peaceful society where people are no longer forced by the laws or economic necessity into monogamous marriages. Women are free; the family is not so much abolished as "enlarged." Citizens unashamedly go about nude, and the population is regulated.

> "And you dare to *regulate* increase! You control it! Your women consent to bear children as they are needed—or refrain!"
>
> "Of course," said Urthred, "Why not?"
>
> "I feared as much," said Father Amerton, and leaning forward he covered his face with his hands, murmuring, "I felt this in the atmosphere! The human stud farm! Refusing to create souls! The *wickedness* of it! Oh, my God!"[60]

His host rationally explains to his twentieth-century visitor that overpopulation led to the wars, crop disasters, and economic depressions that overthrew the old society. Now eugenics hold sway. "The indolent and inferior do not procreate here."[61]

According to C. P. Blacker, president of he Eugenics Society, if birth control were not made available to the masses, Britain would be overrun by the "drunken unemployable" and the "useless parasite." Due to medical advances, many defectives were "now artificially kept alive to perpetuate their kind." Feebleminded women were constantly pregnant. Enlightened workers needed help if they were to ape the middle class's restriction of fertility. Blacker cited a survey that revealed that over half of Saturday night conceptions were undesired.[62] Restriction of fertility, agreed Bertrand Russell, was needed for balanced growth.[63] Blind nature would eventually be replaced by conscious planning. As a result, both the individual and the race would benefit.

Huxley was supposedly inspired by the idea of satirizing Wells's *Men Like Gods* to write *Brave New World*. Huxley had joked, "Mr. H. G. Wells portrays in *Men Like Gods* a race of athletic chemists and mathematical physicists who go about naked and, unlike Mr. Shaw's austerer Ancients, make free love in a rational manner between the experiments."[64] In *Brave New World,* Huxley provides a bleaker view of Malthusianism. Women are either subjected to oophorectomies or, if fertile, use a Malthusian belt with a regulation supply of contraceptives and, if all else fails, turn to abortion centers. Between the ages of twelve and seventeen, girls are subjected to Malthusian drills.[65] "He knows their pleasures," Charlotte Haldane wrote of Huxley's characters "of which the foremost is promiscuous intercourse without fertilization. Here he makes a slight mistake, for no young lady six hundred years hence would wear so primitive a garment as a Malthusian belt stuffed with contraceptives when a periodic injection of suitable hormones would afford her ample protection."[66]

Would, many futurist wondered, women continue to embrace motherhood as their highest calling? As its title suggested, Charlotte Haldane's *Motherhood and Its Enemies* (1927) argued that childbearing was the basis of family life. In the past, the large family and religion were the bulwark of the home. "Modern love," she fretted, was more conditional and therefore more precarious. Birth control allowed women to experiment sexually, and "amateurs" were now elbowing aside paid prostitutes. Such women could not be good wives or mothers since they were likely already infected with cravings for excitement, hatred of the home, and a dislike of children. Indeed, she asserted, such "abnormal women"—thanks to contraceptive protection—were seeking to gratify what were in effect masculine desires.[67] Huxley took repeated potshots at birth control as a fad embraced by unthinking young women trying to appear modern.[68] In a short story, he has an older writer say of a young fan, "You should hear her prattling away about inverts and perverts and birth control—but prattling from unplumbed depths of innocence and practical ignorance. Very queer. And touching too. Much more touching than the old-fashioned innocences of the young creatures who thought babies were brought by storks. Knowing all about love and lust, but in the same way one knows all about quadratic equations."[69]

In *Brave New World,* the goal is a stable population, enforced by the state. This oppressive interpretation of birth control, which presented it as symptomatic of a materialistic age in which the most intimate matters were subjected to planning, was taken up by a number of writers. In *Birth Control and the State: A Plea and a Forecast* (1926), his contribution to the To-Day and To-Morrow series, C. P. Blacker noted that one of the common arguments against birth control was the fear of the bureaucratic surveillance of one's private life.[70] The crassness of planning births was noted even by some of the campaign's supporters. In H. G. Wells's *The Wife of Sir Isaac Harman* (1914), one proponent of the collectivized family form says of the middle class, "But it's not a fine life, it's not a full life, that life in a Neo-Malthusian suburban hutch."[71] Charlotte Haldane agreed that birth control empowered women. Would it be used for good purposes? She worried that being driven by economic necessity to restrict their fertility women were curtailing their own psychological and physiological development. The state could provide mothers with financial support, yet Haldane admitted that there was scant evidence such help increased births.[72] The threat of depopulation was raised by the statistician Enid Charles. She argued that if each woman did not bear at least three children, Britain's population would decline.[73] In an acquisitive society, children were increasingly regarded as luxuries. She suggested that the Soviet Union could be taken as a model of a state providing the sort of social supports that encouraged maternity.

Some offered alternative views of Britain's demographic future. Echoing William Morris, John Gloag imagined an English population happily reduced to the three million citizens who preferred the fresh air of a garden to the stench of industrialism.[74] In Sydney Fowler Wright's *Deluge* (1927), a flood offers the chance of remaking civilization.[75] But most writers of science fiction assumed that the direction in which society was moving could not be altered. Fowler Wright presented birth control and the motorcar as the two key manifestations of modern society's opposition to nature. In his bleak short story "Justice" (1930), he envisaged the restriction of fertility eventually resulting in such a demographic imbalance that by the 1960s the law ceases to penalize the killing of those over the age of sixty-five by motorcar.[76] In her 1926 novel, Muriel Jaeger imagined a man sent into a future of flying machines, euthanasia halls, and economic cooperation. Marriages are more honest, and casual unions are tolerated. Crucially, prosperity is due to the limiting of population.[77] Michael Arlen's *Man's Mortality* (1933), an account of the air battles of the late twentieth century, includes a report of Italy forcing birth control on its Hindu reservation in Africa.[78] And finally some imagined the airplane itself ultimately serving to regulate population. A bizarre account of newborns being thrown out of planes as a form of fertility control was provided by Olaf Stapledon in his futurist fantasy *Last and First Men* (1930). He tells us that far into the future sterilization and contraception are displaced by the ritual of airplane suspension.

> In the life of every individual, flying played a great part. Immediately after birth he was taken up by a priestess of flight and dropped, clinging to a parachute, to be deftly caught upon the wings of his father's plane. This ritual served as a substitute for contraception (forbidden as an interference

with the divine energy); for since in many infants the old simian grasping-instinct was atrophied, a large proportion of the new-born let go and were smashed upon the paternal wings.[79]

Unfortunately for the race, the trial proves to favor the stronger rather than the most intelligent.

What impact would contraception have on the family of the future? The issues of changing gender relationships and access to birth control were necessarily broached by every writer who sought to predict what sorts of new family forms would emerge.[80] Wells, posing as an advocate of free love, in *Anticipations* (1901) imagined that in years to come monogamy would probably still be embraced by the majority, but a certain amount of "regrouping" would also take place as it was illogical to yoke together the unhappy. Nevertheless Wells gallantly granted that it was unfair to cast off a woman whose period of "maximum attractiveness" was short lived.[81] In *Mankind in the Making* (1903), Wells foresaw the old manacles of morality being shed yet sex still being confined by less obvious restraints. For him, the goal of the new freedoms had to be the production of healthy children.[82] In *Days of the Comet* (1906), Wells furthered his attack on the traditional family. Lovemaking in the old society Wells described as crude and unenlightened. Youth were given no sex education. "We were like misguided travelers who had camped in the dry bed of a tropical river. Presently we were knee deep and neck deep in the flood."[83] Struggles for sexual dominance are destructive and humiliating until a mysterious comet's green gas descends on earth and ends paltry passions. The world awakens to the evils of war, materialism, and politics. Sexual jealousy disappears. Youths traditionally brought up to believe that some day they will meet the single person meant for them now know it is not necessarily true. Asking if marriage is not really a form of "extravagant mutual proprietorship," Wells ends the novel with four friends in a sort of group marriage.[84] Such sexual musings were considered shocking when they first appeared. By the 1920s, the younger generation of readers considered Wells's posing as a sex radical inherently fraudulent, given the fact that he always took the male point of view and indeed seemed intent on producing transparent rationalizations of his own erratic sexual behavior.

The most radical of the feminist visions of future family life was Vera Brittain's *Halcyon; or, The Future of Monogamy* (1929).[85] Blurring the line between speculative fiction and predictive social science, Brittain presents a woman professor of moral philosophy at Oxford in the twenty-first century dreaming about the past history of moral progress. In the bad old days of the 1920s, ignorance reigned. "War babies" had been welcomed, but peace babies shunned. Hypocrites defended monogamy while divorce rates climbed; depopulation was lamented by the very people most zealously employing contraceptives. Fortunately the period 1930–75 was an era of sexual reform that resulted in a liberalization of divorce and laws allowing "semi-detached" or intermittent marriages. Genuine monogamy was achieved with the assuring of women's economic independence, the state becoming guardian of all children, the Sexual Instruction Act of 1948 (with mandatory classes), the Trial Marriages Act of 1969, and the Second Contracts Act of 1973 allowing remarriage without divorce during the prolonged absence of a spouse. As a result, the husband no longer felt trapped by the "jealous possessiveness" of a wife confined to the home and hence unable to tolerate his affairs. "Man's need for diversion was indeed comprehensible when his wife resembled at best a well-groomed cow devoid of mental qualities." Wives had once been no more than "carefully-reared animals with whom sex-intercourse was an humiliating concession grudgingly bestowed in return for an economic quid pro quo." Now they were fully rounded individuals who could offer "that variety of experience which married men had once desperately sought from the physical novelties of the mistress or prostitute." Hormonal treatments allowed couples to age symmetrically. Some ectogenesis took place, but just as importantly childbirth was made pleasurable. Moreover "the general standard of intelligence rose rapidly after the sterilization of the unfit became law in 1981," while those who devoted excessive energy to "sex-experiences" lost out to the better disciplined. Even trial marriages declined; most would wait until they were "reasonably sure" before having sex. Promiscuity disappeared, partly due to "minute glandular operations which modify the character of desire." Though occasional open lapses in marital fidelity were allowed, it was understood that they "should be indulged in only as the outcome of mutual agreement between husband and wife." The result was that marriage emerged reformed and refreshed, the supreme expression of love.[86] While others worried about the impact of the machine on morals, Brittain optimistically hailed the "elasticity of domestic life" offered by technological advances. Between 1850 and 1950 many marriages broke down because of the husband having to be away from home for long periods of time. Such long separations due to primitive transport, "so perilous to conjugal fidelity," were no longer necessary in the era of the telephone and air travel. Thanks to the entertainments offered by loudspeakers and "televisors," the home itself became a more welcoming and interesting retreat.[87]

Other optimists chimed in.[88] The socialist physician Eden Paul in *Chronos; or, The Future of the Family* (1930) noted unmarried women now were sexually emancipated because of their access to birth control. Families would become smaller, but the community would serve a role as "over-parent."[89] In his study of the home of the future, Harry Joseph Birnstingl predicted the "uncontrolled breeding among civilized mankind will soon cease altogether," so smaller homes and flats would be needed.[90] The sexologist Norman Haire dared to suggest that in the future, as women grew more prosperous, there would be a probable growth of male prostitution.[91] Bertrand Russell, like most progressives, hailed the coming of smaller families, sex education, and divorce reform. In his eyes, the family at the very least served a function in maintaining the habit of having children. Russell did raise the specter of the father's position diminishing, as the state (as was the case in the Soviet Union) played a greater and greater role in child care. The American behavioral psychologist John B. Watson looked forward to all children being housed in well-run nurseries—not orphanages but wonderful institutions in the countryside where physicians' and nurses' "conditioning hands will not be obtrusive." Watson believed the traditional home would not be missed by the inhabitants of such superior institutions.[92] In contrast, Russell worried about the enormous new powers the state would enjoy once it had the ability to reach children of every class.[93]

Huxley's account of the destruction of the family follows Bertrand Russell's lead. In *Brave New World,* the old family—regarded as obscenely suffocating, inherently pressure packed, and unstable—has been dispensed with. Babies are born in bottles, not to mothers; parenting is replaced by state crèches, and infants are exposed to psychological conditioning. The elite still live in houses, but the masses are confined to barracks. Despite painting such a bleak portrayal of a possible future, Huxley and Russell were attacked by G. E. Newsom, chaplain of King's College, London, as the leaders in the campaign to destroy the existing family. Their "new morality" would result, he predicted, in the state encroaching on the family, in birth control threatening the extinction of the civilized race, in sex and reproduction being split, in fatherhood being made harder to prove, in children being raised in nurseries thereby threatening even motherhood, and in eugenics encouraging only quality breeders.[94] This was a case of Newsom so disliking the message that he shot the messenger.

From the turn of the century onward a number of writers had warned that if socialism ever prevailed, the traditional family structure would be its first casualty. *Brave New World* echoes such works as Horace Wykeham Can Newte's *The Master Beast: Being a True Account of the Ruthless Tyranny Inflicted on the British People by Socialism, A.D. 1888–2020* (1907). As the title suggests, Newte tells of a future socialist society where a ruling class enforces conformity in dress, work, and housing on soulless slaves. Permanent marriage is ended, easy divorce is available, children are raised in state nurseries. Sexual self-denial no longer operates. "Socialism has unhappily removed women's safeguards and inducements to morality, with the result that as a whole (of course, there are exceptions) the sex runs morally amok." Now women, "naked and unashamed," participate in saturnalias. Since they are free of child care and fertility has declined, "caused by the prevailing immorality," it is no surprise that the soulless should prostitute themselves. With no competition the race declines, as made evident by the increased amount of Cockney twang. So weakened by socialism, Britain is finally overrun by "the scourings of Asia and Africa."[95] A similar socialist dystopia was sketched out by F. Britten Austin in *The Red Flag* (1932). After a revolution in the year 1977, numbers replace names and subjects are taught that love is a psychological delusion. Now there is an "entirely free association of the sexes," with no sense of one having a spouse and children.[96] The society's citizens do still pair off; they "register" for a year, and can renew. They give up their babies to the state.[97]

Eugenic concerns implicitly or explicitly framed almost every account of the future of sex, gender, and reproduction. Some writers viewed such a prospect with hope, others with horror. As a large literature has been devoted to the history of eugenics, it is not necessary here to give a detailed account of the movement. What is important to note, however, is that writers from across the political spectrum believed a future society would take measures to assure the reproduction of the healthy and repress the reproduction of the degenerate.[98]

The fact that science-fiction writers exploited eugenic themes is well known; less noted is the fact that many eugenic writings of serious social scientists were patent fantasies. Francis Galton, the creator of eugenics, produced about 1910 a sketch of his utopia, "Kantsaywhere." After his death, the story was apparently censored, and only a few portions appeared in Karl Pearson's biography of Galton. In Galton's tale, the money left by a benefactor is distributed by a college, not to the feeble but to "to help those who were strong by nature to multiply and to be well-nourished." In this community, it is necessary to pass an exam to establish one's fitness to reproduce. Even pronunciation counts. "The 'arry and 'arriet class is wholly unknown in Kantsaywhere."[99] Those who fail are not allowed to breed. If they do not willingly remain celibate, they are forcibly segregated in labor colonies. Patriarchal power is assured, women marrying at about twenty-two, men when established at about thirty. The women are buxom and "seem promising mothers of a noble race." The men are well built and resolute. The elderly Galton obviously savored such a vision. "Both sexes are true to themselves, the women being thoroughly feminine, and I may add, mammalian, and the men being as thoroughly virile."[100] As we have already noted, such scenarios of future worlds or distant utopias where fitness was rewarded and degeneracy repressed predated the science of eugenics. There was a long tradition in the West of passions being regarded as "a threat to regulated social harmony."[101] Ideas about the community's duty to supervise breeding go back as far as Plato's *Republic*.[102] Real possibilities of control only existed from the 1800s onward.[103]

Eugenic views studded the late nineteenth-century utopian writings.[104] In Edward Bellamy's utopia, love matches serve eugenics: "It means that for the first time in human history the principle of sexual selection, with its tendency to preserve and transmit the better types of the race, and let the inferior types drop out, has unhindered operation." Women no longer marry for money. "Every generation is sifted through a little finer mesh than the last. The attributes that human nature admire are preserved, those that repel it are left behind." The result is "race purification."[105] Citizens feel a responsibility to the community to have the best possible marriage. The men not chosen by women are laggards, failures in life. This notion that some control over reproduction would serve to weed out the unfit was almost always assumed in such utopian literature.[106]

In *A Modern Utopia* (1905), Wells portrays a utopian earth controlled by a single world government. Population is its key preoccupation. Misfits—including idiots, invalids, and criminals—are sent, not to lethal chambers, but to prison islands where the sexes are separated to prevent them from breeding, while the better sort are allowed to reproduce. All are cared for by the state, but in return parentage is only allowed the independent. The worst misfits are subjected to "social surgery."[107] In a chapter devoted to "Women in modern utopia," Wells reiterates his argument that motherhood is a racial duty the woman owes the state. She must not be allowed to shirk it. "Our system of morals, therefore, has to make it worth while and honourable to be a mother; it is particularly undesirable that it should be held to be right for a woman of exceptional charm or exceptional cleverness to evade motherhood, unless, perhaps, to become a teacher."[108] As a child-bearer, she in return is entitled to the wages and respect of any other civil servant. Choosing the "profession" of motherhood secures the woman's economic independence. And since the state's only concern is for children it makes no attempt to prevent the dissolution of childless marriages.[109] Discussing "Woman in 2030," Lord Birkenhead likewise predicted that if the family were to survive, limitations would have to be imposed on female intellect. He assumed that maternity would be policed yet said nothing about any changes in men's parenting.[110]

Wells's notion of the endowment of motherhood was a form of positive eugenics. In futurist literature, the fertility of the superior classes was policed by both carrots and sticks. George Godwin mused that the state might interfere in reproduction, by penalizing the fit who refused to breed.[111] The biologist Julian Huxley (Aldous Huxley's brother) noted that the complaint that the best did not breed sufficiently showed the need for tax reform and a racial health service.[112] Birkenhead predicted gene research would lead those considering marriage to question each other's heredity before proposing. "This is the kind of eugenics which will develop in the future, rather than the absurdities of the human stud-farm which so many earnest, and generally unmarried, enthusiasts at present predict."[113] He declared it a "reasonable and even romantic" idea.

Eugenic preoccupations led futurist writers to focus on the issue of differential fertility: the problem of the right types of people having too few children and the wrong kinds too many. "The human race is now passing through a biological crisis unprecedented in the history of life."[114] So asserted Blacker in his 1926 tract arguing in favor of eugenic legislation that would restrict the fertility of the unfit. How would the state carry out such a policy of negative eugenics? Preventing the unfit from marrying was difficult, and segregating them prohibitively expensive. By the twentieth century, however, doctors could provide male sterilizations through the simple means of the vasectomy, thus making the forcible limitation of the fertility of the unfit a real possibility.

Futurist writers found it difficult to avoid the notion that the unproductive would eventually be eliminated. H. G. Wells was particularly prone to harp on the degenerationist threat. In last section of *Anticipations* (1901), Wells hails Malthus as the revolutionary who had ended an era of facile optimism that had believed in biological equality. In the future, experts had to rule, and the unfit had to be eliminated. "It has become apparent that whole masses of human population are, as a whole, inferior in their claim upon the future, to other masses, that they cannot be given opportunities or trusted with power as the superior peoples are trusted, that their characteristic weaknesses are contagious and detrimental in the civilizing fabric, and that their range of incapacity tempts and demoralizes the strong. To give them equality is to sink to their level, to protect and cherish them is to be swamped in their fecundity."[115] In the New Republic, whatever advances the procreation of the fit and retards that of the diseased—even mercy killings—is good. The unfit will be punished for reproducing, and inferior races kept out.[116]

Such ideas were contagious and not surprisingly drew the support of social conservatives. According to Birkenhead, a future society had to have specific types of citizens. "The indiscriminate increase of the most useless type of citizen, accompanied by the voluntary sterilisation of the best type, is the greatest menace which threatens our civilization."[117] In his account of the future of crime, George Godwin protested that sterilization and "painless elimination" were shirked by sentimentalists, yet such policies would protect the greatest number. The degenerate were fertile as weeds, and the taxation costs of supporting such "sub-men" actually had the effect of sterilizing the fit who had to restrict their family size. Though opposed to capital punishment, Godwin applauded the elimination of threats to the race. "In the future," he predicted, "we shall probably regard it as a meritorious act on the part of a mother that she destroys at birth, say, a Mongol idiot."[118]

Among the bleakest eugenic accounts of the future was the philosopher F. C. S. Schiller's *Tantalus; or, The Future of Man* (1924), which claimed that evolution was being reversed, the higher classes failing to reproduce, while the feebleminded and casual laborers spawned huge broods. The dysgenic effect was compounded by the government preventing the poor from having access to birth control and by medicine and philanthropy thoughtlessly engaging in "baby-saving," that is, rescuing infants who should not have survived. If such actions continued, Schiller foresaw the race likely to peter out "in an overwhelming flood of feeble-mindedness." Eugenic sterilization was needed. Countering Russell's worries about the tyranny of science, Schiller asserted that such interventions would begin cautiously. "We start with a pretty shrewd suspicion that certain types, say the feeble-minded, the sickly, the insane, are undesirable, and that no good can come from coddling and cultivating them."[119] At the same time, to restart evolution, various "experiments" would be undertaken to increase the fertility of the superior classes. Norman Haire, in his account of the future of marriage, made much the same arguments as Schiller. Haire described the possibility of sterilizations being carried out by x-rays, superheating of the gonads, or some form of immunization. He foresaw the unhealthy being ordered to use contraceptives. If they disobeyed, the offspring would be aborted or the decision made after birth if it should be allowed to live. Indeed he foresaw in a future society the establishment of boards that would decide on eugenic grounds on "painlessly destroying persons who are a menace to it."[120]

Science-fiction writers found it difficult to dream up scenarios bleaker than those produced by the eugenicists, but they tried. Olaf Stapledon, in *Last and First Men: A Story of the Near and Far Future* (1931), argued that with humanitarianism in vogue the unsound reproduced rottenness. In Frederick le Gros Clarke's uninspiring novel *Between Two Men* (1935), an embryologist declares that the species is unstable and wearing out; a new one will emerge, the Übermensch will appear, and the unfit perish.[121]

Progressive writers were aware that the specter of biological degeneration was exploited by social conservatives. Vernon Lee noted that the public was bombarded daily with reports of new threats, and not just by blacks, Jews, Bolsheviks, and fascists. "There is Peril from the multiplication of Idiots and the multiplication of Supermen; Peril from depopulation and Peril from overpopulation, from unsexed women and over-sexed women; Peril from over-much altruism, and Peril from insufficient altruism."[122] The fate that awaited a society that fell to the eugenicists was presented by Sydney Fowler Wright in his 1929 short story "PN 40." He portrayed a repressive "Eugenic Era" in which the unfit are forbidden to marry while a board of allocation arranges the polygamous marriages of the fit. Women are married off early and rewarded with a medal for each child they have. A young woman called PN 40 is drawn to 48 VC, but as he is allowed only one-quarter of a wife, the couples' happiness depends on their escaping to Brazil.[123]

Despite sharing Wright's distaste for eugenics, many writers on the left found it difficult not to believe that science would eventually solve society's social ills. Dora Russell insisted that workers would be just as good as the middle classes if they had the same food and training. Poor conditions led to feebleness of mind. Accordingly while arguing that workers needed access to birth control, she insisted that the large working-class family did not pose a threat.[124] Enid Charles agreed that the alarmist assertions of eugenists were unproven. They failed to see that most defectives were born to normal parents, so it was quite wrong to talk about an increase in defective stock. Yet Charles still spoke of the "quality" of the population and conceded that the encouragement of "socially valuable" stocks was important.[125] Charlotte Haldane asserted that intelligence was inherited, but that it needed a good environment if it were to flourish. It made better sense therefore to seek to improve the environment than follow the policies espoused by eugenic cranks with their obvious class and race prejudices. Yet she could at the same time argue that the eugenicists were right in supporting the sterilization of the handicapped: "As far as I can see, there is no reason whatever why such people should be allowed to reproduce, though all sexual intercourse need not necessarily be denied them if the community does not suffer by it."[126]

Her husband, the Cambridge scientist J. B. S. Haldane, presented the spread of contraceptives as an example of the practical application of biology gone stupidly awry: "The result is that the only means available of regulating population, a means which might be of immense social value if it were directed, is allowed to be used in the most haphazard and unsatisfactory manner."[127] Turning to sterilization, Haldane ridiculed the eugenic "off-with-your cock-brigade" and the Nazi fanatics, but still had faith in the emergence of an ultimately benign eugenics.[128] Scientific reasoning had to win out eventually. Bertrand Russell took a similar stance, asserting that, "government opposition to birth-control propaganda gives a biological advantage to stupidity, since it is chiefly stupid people whom governments succeed in keeping in ignorance. Before long, birth-control may become nearly universal among the white races; it will then not deteriorate their quality, but only diminish their numbers, at a time when uncivilized races are still prolific and are preserved from a high death-rate by white science."[129] Russell accepted that eventually sterilizations would have a role to play. "We may perhaps assume that, if people grow less superstitious, governments will acquire the right to sterilize those who are not desirable as parents." One could start with imbeciles, then epileptics, consumptives, dipsomaniacs, and finally "all who fail to pass the usual school examinations."[130] Though he conceded that a few geniuses would be lost, he believed average intelligence would improve.

Julian Huxley was skeptical of many of the eugenicists' claims, and noted that environmental influences could mask genetic variations, but he too accepted the argument that the lower classes were too fertile. They needed birth control; unless they had small families, they should be prevented from having easy access to relief or hospitals. Long unemployment he suggested be made grounds for sterilization.[131] J. D. Bernal, another leftist scientist disdainful of the eugenicists and faddists who judged social worth on externals, nevertheless accepted the need to control population. He foresaw that "the better organized beings will be obliged in self-defense to reduce the numbers of the others, until they are no longer seriously inconvenienced by them."[132] Indeed he playfully speculated that the day might come when the enlightened moved to other planets leaving the earth as a sort of zoo for misfits.[133] All these writers were clearly motivated by both a faith in science and a fear of the masses.

It was in this cultural context that some sketched out the possibility of ectogenesis.[134] The necessity of attempting to restrict the fertility of some citizens could be preempted if the reproduction of all was controlled by the state. This was Huxley's vision of babies in bottles, a logical if outlandish response to on the one hand a series of threats to reproduction including undisciplined women and emasculated men and on the other hand the prospect of racial improvements promised by eugenicists, biologists, and medical scientists. The fascination with the idea of a scientist creating life can be traced at least as far back as Mary Shelley's *Frankenstein* (1817). Fact began to catch up with fantasy when in Britain in the 1890s the Cambridge scientist Walter Heape carried out the artificial insemination of rabbits and the transfer of their fertilized eggs.[135] At about the same time in America, the press portrayed Jacques Loeb as a "Dr. Frankenstein" when news leaked out of his work on "artificial parthenogenesis." Though Loeb's notoriety was won by his simply stimulating sea urchin eggs to begin development, an 1899 account presented him as attempting "to create Life in a Test Tube" while another was titled the "Creation of Life."[136] Similar powers were attributed to the French researcher Alexis Carrel who in the early twentieth century moved from cell research and organ transplants to preserving grafts.[137]

In a 1914 lecture, J. B. S. Haldane captured the optimism of biologists, aware that they were participating in a radical new shift in the relationships between humans, plants, and animals.[138] This lecture was the basis of a small book, *Daedalus; or, Science and the Future* (1924), in which he asserted that the current progress of medicine promised to rival in its social impact that of the Industrial Revolution. The Great War had raised the threat that science might destroy man or make him a "mere parasite of machinery, an appendage to the reproductive system of huge and complicated engines."[139] As scientific progress could not be stopped, the challenge, according to Haldane, was how best to apply biology to human life. Most foresaw only a few changes in medicine and eugenics, but in *Daedalus* Haldane set his sights higher. He imagines an undergraduate 150 years hence, looking back on a world transformed by remarkable scientific achievements. The first ectogenetic child is born in 1951—the result of Heape's transfer of rabbit embryos, Haldane's 1925 work on embryonic rats, and the ability of scientists of the 1950s to keep ovaries alive, harvest their eggs, and fertilize them.[140] At midcentury, sexual love and reproduction are finally split, with only 30 percent of babies born of woman.[141] Such reproduction by design proves to have obvious social benefits. "Had it not been for ectogenesis," we are told, "there can be little doubt that civilisation would have collapsed within a measurable time owing to the greater fertility of the less desirable members of the population in almost all countries."[142] In short, thanks to science the race is both changed and improved.

In producing his account of how test-tube babies would soon be produced, Haldane was provocatively portraying the biologist as the most romantic and most blasphemous of figures. Every new invention or discovery—even milking a cow—had been initially damned as a "perversion," as "indecent and unnatural," he noted, but eventually was normalized. So too, he asserted, thanks to the biological sciences, sex and the family would inevitably be reformulated. Even before the appearance of *Daedalus*, Haldane's disturbing views were being publicized. In his comic novel *Crome Yellow* (1921), Aldous Huxley has Scogans—a character based on Haldane—declare his unbounded faith in scientific progress:

> With the gramophone, the cinema, and the automatic pistol, the goddess of Applied Science has presented the world with another gift, more precious

> even than these—the means of dissociating love from propagation. Eros, for those who wish it, is now an entirely free god; his deplorable associations with Lucina may be broken at will. In the course of the next few centuries, who knows? The world may see a more complete severance. . . . An impersonal generation will take the place of Nature's hideous system. In vast state incubators, rows upon rows of gravid bottles will supply the world with the population it requires. The family system will disappear; society, sapped at its very base, will have to find new foundations; and Eros, beautifully and irresponsibly free, will flit like a gay butterfly from flower to flower through a sunlit world.[143]

Haldane's sketch of ectogenesis inspired a number of more serious writers including leftists such as the crystallographer J. D. Bernal, the sexologist Norman Haire, and the geneticist H. J. Muller.[144] Haldane's vision also met with a range of critical responses. Skeptics included such futurists as E. E. Fournier d'Albe. While noting Haldane's ectogenesis, he held that though in coming decades the forms of marriage might change, parenthood would not.[145]

In *Icarus; or, The Future of Science* (1924), Bertrand Russell, worried that biological discoveries would be used by the elite to control the masses, questioned Haldane's optimism. Russell was of two minds concerning the control of reproduction. Sheep were scientifically bred, and the idea of applying such measures to humans had its appeal. "The study of heredity may in time make eugenics an exact science, and perhaps we shall in a later age be able to determine at will the sex of our children."[146] However, such interventions could have disastrous unintended consequence, such as a surplus of male babies. Nevertheless Russell accepted that "it may easily be that within the next hundred years the sciences of heredity and bio-chemistry will have made such strides as to make possible the breeding of a race which everybody would admit to be superior to that now existing."[147] Though he found the idea repugnant, he also conceded that only a small percentage of breeders would be necessary. In theory, one could have a society of Edisons with the strength of prizefighters, but Russell warned that once science knew more about reproduction, it would interfere more in family life than religion had ever done. "Nevertheless," he concluded, "if there is to be a tyranny, it is better that it should be scientific."[148]

Russell pushed his message further in *The Scientific Outlook* (1931), noting that the Soviet Union was demonstrating how malleable people were. Reproduction was now controlled, and sex selection was inevitable. With the increased policing of reproduction, he foresaw the possibility of only about 5 percent of men and 25 percent of women being selected as breeders; the rest could be sterilized. Though the breeders would each have eight to nine children, they would have no other work. He mused that artificial impregnation might be employed to such an extent that eventually it would be not "ladylike" to be impregnated in "natural manner." And if mothering was largely replaced by incubators and nurses, fatherhood would all but disappear.[149] Was this socially desirable? Bertrand Russell foresaw eugenics remaking maternity. "Moral standards may alter so as to make it possible for one man to be the sire of a vast progeny by many different mothers."[150] Yet Russell continued to voice his concern that any system run by officials and doctors would unlikely be wiser than nature.[151]

Julian Huxley turned to fiction to warn of the dangers of scientific hubris. In his story "The Tissue Culture King" (1926), he tells of a captured medical scientist who works for an African tribal chief by exploiting tissue cultures, experimental embryology, endocrine treatments, and parthenogenesis to produce freakish men and animals. He succeeds in reproducing birds and reptiles by parthenogenesis, but escapes before being forced to create humans.[152] In presenting a frightening vision of the excesses to which the reproductive sciences might lead, Huxley echoed H. G. Wells's cautionary tale *The Island of Dr. Moreau*.

A misogynist like the conservative social critic Anthony Ludovici who paraded his hostility to the "artificial"—including spectacles, forceps, and baby food—was predictably hostile to Haldane's vision. According to Ludovici, women and scientists were alike in despising the body. Doctors already interfered in the birthing process. In the future, he foresaw politically empowered feminists, not satisfied with having pushed back the age of consent to thirty-five or forty, happily embracing ectogenesis. "When once artificial impregnation is an every-day occurrence, a Parliament of women will doubtless pass legislation to make it illegal for any man to procreate a child naturally, if it is the wife's desire to have one by the intermediation of science."[153] First, the naturally fertilized will be despised. Next, extracorporeal gestation (following Carrel's work) via a cow or ass will be possible. Children, Ludovici predicted, would be "grown" in incubators by the local council. Only a small number of men would be needed for fertilization; most would be superfluous. And once women were in power, he asserted, "rape will be punished brutally probably by means of emasculations; and men of vigorous sexuality will be eliminated in order to make way for a generation of low-sexed, meek, and sequacious lackeys."[154] Owen Johnson's novel *The Coming of the Amazons* (1931) aped Ludovici's warning in portraying a society in which women were flat chested (as they no longer nursed) and bald (because they were so intelligent). The state, following eugenic guidelines, created specific castes, forbade families, made birth control obligatory, and determined the sex of children who were raised in nurseries. Women, being in power, made the sexual advances, and men were eliminated after their services as breeders were concluded. The story ends with "John," the narrator, futilely trying to rouse the men to rebel.[155]

Birkenhead concurred that in a society in which marriage and reproduction were split, the "feminist dream of equality" would finally be achieved. Freed of childbearing, women could decide on whether or not to adopt ectogenetic babies. The foster mother would be prompted by science to produce milk and emotions "as though she had herself borne it."[156] The woman with a genius for mothering might adopt a huge brood; others, if they chose a career, might have none. Birkenhead argued that as ectogenetic birthing could trump eugenic concerns about breeding, it was likely to occur. The state could then determine the choice of parents. "The Cabinet of the future could breed a nation of industrious dullards, or leaven the population with fifty thousand irresponsible, if gifted, mural painters."[157] It would naturally breed those it "needs." Workers would not be slaves since like worker bees their happiness would be found in working. In fact, in a machine age there would be little work to do, and the challenge (as in *Brave New World*) would be how to fill leisure time.

What did women think of such scenarios? Although as a feminist she was exactly the sort of woman Ludovici abhorred, Charlotte Haldane expressed similar eugenic concerns in her contribution to the discussion. In her futurist novel *Man's World* (1926), white males control women's sex and reproduction. The twenty-first-century government decides how many children will be born. Geneticists apply a method of sex selection while Motherhood Councils are in charge of matings. Youths are married off early; women either become mothers or are "immunized" to protect the race from haphazard breeding. In the old days, all women were forced into motherhood, though many became neurotic as a result. Now only the healthy are destined for motherhood; the neuters are not. Moreover the mothers of the "white race" are not expected to be full-time wives. They have only one primary task, that of "race-production," which is carried out in maternal settlements. Birth control has been followed by "sex control," meaning women now impose their will on men. This society of mothers has its mechanical slaves, neuters, and, as in Rome, "entertainers" to deal with men's lust. Haldane thus foresees women outgrowing the Christian family, and she has a male scientist predict that ultimately ectogenesis will spare most women even childbearing.[158]

Like Charlotte Haldane, other feminists were ambivalent about talk of scientists assuming a greater control of reproduction.[159] In her discussion of threats to population growth, Enid Charles cited J. B. S. Haldane's discussion of test-tube babies and warned that his vision was closer to being realized than some might think. Rabbit embryos were already being cultured in vitro and implanted. Sex determination and artificial insemination would be soon possible for humans. It was even possible to envisage parthenogenesis, in which males would not be needed and twins rather than single babies might be preferred.[160] In *Halcyon; or, The Future of Monogamy*, Vera Brittain agreed that ectogenesis would eventually take place but felt it would be limited to prevent the "demolition of the human race."[161]

Huxley's *Brave New World* described a society that pushed to its logical conclusion the eugenic desire to produce standardized humans. It is necessarily inegalitarian, likened by the Controller Mustapha Mond to an iceberg with eight-ninths of its mass below the waterline. In this world, class becomes biology, with each individual designed for their particular task. Repetitive factory work is impossible for the intelligent Alpha but fine for the moronic Epsilon. Accordingly, the development of some test-tube babies is subjected to cell division and retardation to provide standardized workers or "identical machines."[162] The same argument against social leveling was made by Blacker. There was no point in educating those who had to do the dirty work: "It is doubtful whether the possession of a very high degree of intelligence would make such workers happier or more efficient."[163]

Such a vision had its critics. The leftist biochemist and sinologist Joseph Needham protested that Huxley revealed himself to be too Platonic in his hankering after a slave society. Instead of having to rely on morons, a truly advanced community would make work so interesting that it would need the more intelligent.[164] Though not opposed in principle to the creation of an ectogenetic labor force, Birkenhead predicted it would not be necessary as "production will become so cheap, and, barring political or international upheavals, wealth will accumulate to such an extent, that the ectogenetic Robot will never be needed."[165] References to ectogenesis naturally enough cropped up in subsequent science-fiction works. In the year 20,000 AD, society has its "breeding racks" of babies in bottles in Laurence Manning's *The Man Who Awoke* (1933).[166] In *Odd John* (1935), Olaf Stapledon foresaw colonies of "supernormals" reserving sex for love and reproducing via ectogenetically fertilized ova.[167]

Why did these discussions of ectogenesis emerge when they did? They were obviously fueled by reports of breakthroughs being made in the biological sciences, but more important, they were also a response to worrying shifts in gender relations and apparent threats to population. Such works bear out Fredric

Jameson's assertion that science fiction's chief ability is "to demonstrate and to dramatize our incapacity to imagine the future."[168] George Woodcock agrees that "though *Brave New World* is projected on to the screen of the future, it is derived almost entirely from tendencies which Huxley observed with alarm and distrust in the world around him."[169]

If Huxley's portrayal of reproduction in the future was, because of his wit and stylistic accomplishments, the most memorable, it was not all that original. Critics claimed he was simply a popularizer of ideas. "On inspection," sniffed Q. D. Leavis, "his learning is found to be painlessly acquired like his information from such obvious sources as encyclopeadias, the scientific best-sellers, the current popular sociological, psychological, anthropological, etc works." Of course, for that very reason Huxley is a good guide to his age, and even Leavis conceded Huxley had a "flair for embodying the Zeitgeist."[170] In *Brave New World*, an American reviewer noted, Huxley held up a mirror to the age: "the standardized world we are facing and fearing, a world made safe by science and psychology as they are promoted by the eugenicist and the behaviourist."[171]

For the purposes of this study, Huxley is important in simply being the best known of the many writers of his generation who, in writing about the future of reproduction, were torn between an admiration for planning and a distrust of the powers of science. Such works are obviously of little value if judged on their authors' abilities to predict what would happen in the year 1977 or 2020 or 2073 or 20,000. The interest of these books really lies in revealing how, in purportedly anticipating a world to come, cultural commentators transparently drew on contemporary worries about the impact a machine-dominated society was already having on sex, gender, and fertility.

CHAPTER 2

"What is better, a car or a wife?"

Automobiles and Modern Bodies

In the previous chapter, we saw that many British writers who speculated on the future were preoccupied by the notion that people would become more like machines, and machines more like people. For the early twentieth-century reading public, one machine held a particular fascination—the motorcar. *Brave New World* is set in the year A.F. 632, that is to say 632 years After Ford. Huxley imagines a world that looks back on the car manufacturer as a sort of god. Instead of a crucifix, its inhabitants wear a "T" (in remembrance of the Model T) and swear by "Our Ford." However, he is venerated chiefly as a prophet of mass production, not as a producer of cars. Motorcars no longer exist; the lower castes are transported by monorail tramcars and members of the elite by rocket planes and helicopters. They teach their children to repeat, "I do love flying. I do love flying."[1]

Other futurists like H. G. Wells similarly predicted a coming age of airplanes and high-speed trains and said little about the automobile. In fact, the full social and cultural impact of the motorcar was in the interwar years only beginning to be appreciated. It could be an invaluable servant, stated *The Times* in a 1928 editorial, but it could turn into a "tyrannous and destructive Robot."[2] For our purposes, what is particularly striking is that social commentators were voicing the claim that more than a mere means of transport, the automobile was intimately entangled in changing gender relationships and ultimately in reproduction itself. Accordingly, as a way of understanding what British culture made of the human/machine relationship, this chapter investigates what it said about sex and cars.

No single machine was more representative of twentieth-century modernity than the motorcar. Its enormous cultural significance is demonstrated by the fact that, as Allen Samuels states, the British used it almost immediately to help them "describe, understand and reflect" on their society.[3] Any discussion of the car appeared to lead inexorably to observations on its reflection of and probable impact on class and gender relations. But what is even more noteworthy is that one can detect in the works of many middle-class writers the belief that the automobile played a central social and cultural role in helping to organize sexuality. As such, it could either help or hinder reproduction. Of course, the argument that the automobile enhanced the power of heterosexual males is a familiar one. But a survey of representations of automobility in the first half of the twentieth century demonstrates that the story is far more complicated. Commentators believed that mechanized transport could have an impact on sexuality in a variety of often contradictory ways—the car was credited with undermining virility and restoring it, with reinforcing gender boundaries and eroding them. How could a machine have such apparently incompatible consequences? To provide some answers, the first part of this chapter will primarily exploit elite and popular fictional representations to demonstrate the extent to which motorcars were viewed as reflecting their owners' emotional and sexual desires. The goal is to appreciate why, in short, automobiles were viewed as sexed.[4] The second portion of the chapter will draw more on nonfictional sources to make the less familiar argument that if commentators commonly anthropomorphized the car, more disturbingly, they came to portray sex as "automobilic." The overarching argument is that the motorcar was always more than a mere means of transport; it served as a vehicle for the broaching of some of the most intimate areas of private life.

In what ways did the car mirror the erotic interests of men, women, and youths? From the very beginning the insightful portrayed car ownership and driving as forms of masculine sexual display. The motorcar's first attraction was its speed. Speed, Aldous Huxley declared, "provides the one genuinely modern pleasure." He was referring in particular to the velocity of the car. Swiftness had perhaps always been admired and trains and planes now offered their passengers rapid transport, but the automobile under the command of the individual driver was unsurpassed as an "intoxicating speed-purveyor."[5] The poet Osbert

Sitwell agreed that due to the arrival of the motorcar in the early 1900s "there was the awareness of speed itself, and the rapid thinking that must accompany it, a new alertness . . . all these physical impressions, so small in themselves, went to form a sum of feeling new in its kind and never before experienced. . . . The open car belonged to that day. No other generation had been able to speed into the sunset."[6]

The world of speed was not welcomed by all. A racist crank like R. Austin Freeman voiced the alarm that society was in danger of being enslaved to machines. Everyone had access to "power," and the result, he warned, was degeneration, racial deterioration, and national decline. The railways had blighted the countryside, and now in the 1920s one witnessed the emergence of "a locomotive type of man," the restless, impatient "hustler," the motorists addicted to the "speed-habit."[7] The irony, claimed Freeman, was that the new transport did not save time; it absorbed more of it. The philosopher C. E. M. Joad likewise fulminated against the "fantasy world of speed and physical sensation."[8] Some of the most savage attacks on the automobile were penned by the Social Darwinist, misogynist, and anti-Semite Anthony Ludovici. "The thrill of speed which, I believe, could be shown to be due largely to sadistic impulses, the feeling of power which speed imparts and, above all, the sense of superiority associated not only with higher speeds of travel but also, through the association of at least owner-driving with the pecuniary prestige of such means of travel—all these factors combine to make the owner-driver deaf to any humanitarian or civilized appeal."[9] Pedestrians, claimed Ludovici, were now as savagely treated as slaves and horses once were.[10]

Whatever their views on speed, social commentators tended to agree that possession of a car in many obvious ways reinforced the power of its owner—usually a male. Sociologists have observed that for most young men masculinity means taking up space by gesture, movement, or noise. The car multiplies such appropriations.[11] The possession of a motorcar, writers asserted, extended a man's authority and the submission, sexual and otherwise, of others. "The vehicle, so modern and of its time," recalled Osbert Sitwell, "induced in the young man a sense of being heir of all the ages, lord of all he passed by. For the sense of speed flatters the sense of power, raising the rich, and even the humbler lorry-driver, to a new and god-like level."[12] Some writers went so far as to present the motorcar as the male's doppelganger. E. M. Forster in *Howards End* (1910) refers to the "throbbing stinking car" of the Wilcox men violating the calm of the countryside.[13] In *The Wasteland* (1922), T. S. Eliot likens Sweeny to an automobile when he sets off to seduce a typist:

At the violet hour, when the eyes and back
Turn upward from the desk, when the human engine waits
Like a taxi throbbing waiting.[14]

Others regarded the car as a seductress, its chief victim being the man and his enslavement likened to a sort of sexual addiction. Kenneth Grahame in *The Wind in the Willows* (1908) was probably the first to describe the roadster as sexual surrogate. When Toad's friends deprive him of his car and keep him home, Grahame daringly portrays his hero as resorting to a sort of self-abuse. "When his violent paroxysms possessed him he would arrange bedroom chairs in rude resemblance of a motorcar and would crouch on the foremost of them, bent forward and stare fixedly ahead, making uncouth and ghastly noises, till the climax was reached, when, turning a complete somersault, he would lie prostrate amidst the ruins of the chairs, apparently completely satisfied for the moment."[15] Sitting in a puddle and croaking is what lovesick amphibians do, and Toad clearly needs a car as a sort of sexual fix. In *Vile Bodies* (1930), Evelyn Waugh asserted that for some the automobile provided simple locomotion—the wealthy had the Hispano-Suiza or Rolls Royce or Daimler; the humble the Austin Seven. "Not so are the *real* cars, that become masters of men; those vital creations of metal who exist solely for their own propulsion through space, for whom their drivers, clinging precariously at the steering wheel, are as important as his stenographer to a stockbroker."[16]

In the twentieth century, the challenge of mastering an automobile provided a new way for a man to prove his masculinity. To be "in the driver's seat" was to be in control, to be independent, to be manly. The car was redolent of modern manhood, exalting as it did the individualism of the driver as opposed to the cowed conformity of the railway passenger. Accordingly, literary historians have remarked on the "obsessively doted-on aids to speed, a motorbike, a racing-car, or airplane" found in so many interwar fictions.[17] In John C. Moore's novel *The Walls Are Down* (1933), the hero is described as looking half-asleep most of the time "except when he was writing a leader, or climbing a rock face, or driving a car. Then he became taut and eager suddenly: a fighting-man all over." Audrey prefers such men of action to Bloomsbury intellectual types and is pleased that Harry races at Brooklands.[18] Similarly the bright young people who figure in R. Denne Waterhouse's *Week-End Ticket* (1934) are rejuvenated when driving. At a dinner party, they act out a mock Brooklands race on a restaurant table, the guests aping the sounds of speeding cars.[19]

In combining speed, risk, and danger, the motorcar produced a masculine aphrodisiac; or at least it was portrayed as such by both psychologists and novelists.

Freud did not address the issue of cars per se, but in writing about infantile sexuality he noted children's excitation and satisfaction was obvious in the pleasure they took in being swung or thrown about. This led him to assert, "The shaking produced by driving in carriages and later by railway travel exercises such a fascinating effect upon older children that every boy, at any rate, has at one time or other in his life wanted to be an engine driver or a coachman."[20] The result was "a compulsive link" between rail travel and sexuality. Psychoanalysts extended the scenario to include motoring. Adler referred to speeding in a car as sexually exciting the man; the woman in the passenger seat would enjoy a vicarious thrill.[21] For Michael Balint, each new adventure, be it with a new car or new sexual partner, provided the thrill of a "virgin" experience.[22]

Cars were billed by advertisers as providing men with a reassurance of their virility. The unspoken message was that fears of impotence could be compensated for by the speed of a car's acceleration. Now even the weakling could be potent, if he possessed a car. He might not boast directly of himself but could legitimately laud the powers of his motor.[23] In novels, forceful driving was presented as a metaphor for potency.[24] In Aldous Huxley's *Those Barren Leaves* (1925), Lord Hovenden, though inherently meek, while driving at high speed becomes daring and proposes marriage.[25] In *Point Counter Point* (1928), Huxley presents Everard Webley, the fascist, carrying off Elinor Charles for a tryst in his car and notes the "furiousness" of his driving. "He was such a good power-lover" who cherished the "power and sense of superiority" given by his car. "It was a powerful machine (for Everard was a lover of furious driving) . . . he shot off with violent impetuosity." Elinor finds driving with him "almost too exciting."[26] Similarly in *Hangover Square* (1941), Patrick Hamilton sketched out George's drunken and hopeless infatuation with the mercenary Netta of Earl's Court. Netta favors the vicious fascist Peter because of his criminal past. He was known "for killing a pedestrian with his car while drunk, and this she liked, this stimulated her."[27]

Automobile advertisements exploited male fantasies of gaining increased power, and male fiction writers carried on the message by presenting men successfully employing cars to seduce women.[28] In *Howards End,* though at first shocked by the middle-class crassness of Mr. Wilcox and his family, the bohemian Helen Schlegel "leant back luxuriously among the cushions of his motorcar" and allowed herself to be charmed.[29] In Gilbert Frankau's six-hundred-page tear-jerker *Farewell Romance* (1936), David takes Judith out into the country for a joy ride. She shares the married man's enthusiasm for speed. "He put his foot down again; and the barrier between them seemed to fall."[30] D. H. Lawrence has the young woman in *The Virgin and the Gypsy* (1930) being impressed by "such a smart young officer, awfully good class, so calm and amazing with a motorcar, and quite a champion swimmer."[31] Birkin's seduction of Ursula in Lawrence's *Women in Love* (1920) begins by her being at first frightened and then intoxicated by his dangerous driving. "Driving in a motorcar excited her, she forgot his presence even."[32] Finally, in his short story "In Love," Lawrence pushed the identification of the women with the car to ludicrous lengths. A female character muses about her car-loving fiancé:

> And surely she, Hester, was as complicated as a motorcar! Surely she had as many little valves and magnetos and accelerators and all the rest of it, to her make-up! If only he would try to handle *her* as carefully as he handled his car! She needed starting, as badly as ever any automobile did. Even if a car had a self-starter, the man had to give it the right twist. Hester felt she would need a lot of cranking up, if ever she was to start off in the matrimonial road with Joe. And he, the fool, just sat in a motionless car and pretended he was making heaven knows how many miles an hour.[33]

In short, owning and controlling a car should have led on to the man's possession and pleasuring of the woman.[34]

The advent of the car was commonly associated with the escape and sexual liberation of not just men in general, but of young men in particular. Osbert Sitwell stressed that his was "the first generation in which young men were allowed to take their sweethearts for drives—only the fastest of fast actresses had ridden in tandems. . . . They would sit together, the two of them, the man at the wheel, the girl beside him, their hair blown back from their temples, their features sculptured by the wind, their bodies and limbs shaped and carved by it continually under their clothes, so they enjoyed a new physical sensation, comparable to swimming; except that here the element was speed, not water."[35] In the United States, the car was described by Dorothy Bromley and Florence Britten in *Youth and Sex* (1938) as "an incredible engine of escape."[36] In their classic work *Middletown: A Study in American Culture* (1929), Robert S. Lynd and Helen Merrell Lynd observed that the car had extended the mobility of youth and, along with movies, allowed boys and girls to sit together in the dark and stay up till midnight. They cited a juvenile court judge who went so far as to assert that "the automobile has become a house of prostitution on wheels."[37] To back up such moralizing, the authors cited the fact that in a 1924 sample of thirty girls charged with sex crimes, nineteen had been committed in a car. Wayward boys in contrast, increasingly bored by industrial work, were understandably tempted by the appeal of "speed" to steal cars.[38]

British commentators were much impressed by the mobility of American youth. Bertrand Russell reported in *Marriage and Morals* (1929) that thanks to the automobile it was easy for young American women "to escape with a lover from the eyes of parents and neighbours."[39] In Britain, youth had less access to cars than their American counterparts. The Oxbridge colleges tried to limit motorization. "What most concerned the proctors," writes Paul Deslandes, "were the erotic possibilities inherent in car ownership."[40] The mobility and privacy offered by the car would allow the crossing of geographical and class boundaries. And so the same sexual script emerged in Britain and the United States. Getting into a man's car was taken as tantamount to consenting to sex. The music hall comedic singer Vesta Victoria slipped a few suggestive sexual innuendos in "Through Riding on a Motorcar" (1903), lamenting that as a result of going for a drive:

> I've lost no end of things
> Don't know where I are.

And as a result of a crash her beau Jim has a broken limb so she hasn't had her honeymoon yet. The religious did not view joy riding as a laughing matter. In the 1920s, Irish bishops and priests "were much concerned at 'the misuse of . . . motorcars for luring girls' at country dances."[41] A decade later Mrs. Florence Booth, widow of General Bramwell Booth of the Salvation Army, complained to the Home Office that the country lanes, especially near Barnett and Potters Bar, were "infested with motorcars drawn up by the roadside . . . virtually brothels on wheels."[42]

In the interwar period, young people in Britain were in reality far more likely to court and travel on trams, trains, and buses than in cars. And in contrast with American writers, British authors tended to present gloomier portrayals of cars and youthful sex. Mention of the "back seat" conjured up images of seduction or worse. Waugh ended his novel *Vile Bodies* with the feckless Charity in the back seat of a Daimler with a general.[43] In Graham Greene's *Brighton Rock* (1938), a boy crudely requests sex with the command "Come out to the cars." But when the girl climbs in the back seat, hikes up her skirt, and asks, "You've got the doings, haven't you?" he panics.[44] Graham Greene's "A Drive in the Country" (1937), begins with the prospect of a car providing young lovers with a means of escape but concludes with a disastrous dash for freedom, the boy committing suicide and the girl returning home.[45]

Cars, in providing mobility and privacy, were naturally enough believed to facilitate men's extramarital affairs.[46] In a 1926 play about divorce and adultery, a lawyer asserts that the changes in contemporary morality are due not to the suffragists' successes but to new opportunities offered by technologies like the automobile. "St. Paul and John Knox have proved no match for Dunlop and Daimler."[47] Brighton, within easy motoring distance of London, had a raffish reputation for naughty weekends that resulted in the collection of evidence for the divorce courts. In *Brighton Rock* (1938), Graham Greene observes, "young men kept on arriving in huge motoring coats accompanied by small tinted creatures. . . . They looked at nobody, sweeping through the lounge as they had swept in racing models down the Brighton Road, ending on high stools in the American bar."[48] Such roadsters reek of sex. "A scarlet racing model, a tiny rakish car which carried about it the atmosphere of innumerable road-houses, of totsies gathered round swimming pools, of furtive encounters in by-lanes off the Great North Road."[49] The automobile thus offered men both new ways of being domestically domineering and new ways of escaping the obligations of home.

But was the lure of the motorcar used by men only to seduce women? Sarah S. Jain argues that the automobile—in playing the role of "a male prosthetic to get the girl to impress the guys"—can also promote homosocial relations.[50] Given that more than one interwar commentator implied that a woman's excessive interest in cars could be taken as a symptom of lesbianism, it is striking that few pointed out the corollary—that an obvious homoerotic current ran though much of men's passion for driving. In *Brideshead Revisited* (1944), Evelyn Waugh does portray Sebastian employing his two-seater to seduce Charles. "I must say the whole of Oxford has become *most* peculiar suddenly. Last night it was pullulating with women. You're to come away at once, out of danger. I've got a motorcar and a basket of strawberries and a bottle of Chateau Peyraguey—which isn't a wine you've ever tasted, so don't pretend."[51] Most novelists skirted the topic, but some sense of the role that motorcars played in enabling homosexual encounters is found in court reports and newspaper coverage of blackmail trials. For example, in 1932 a retired businessman told the police that he had met at the bar of the Palladium Music Hall a young fellow called Alfred Bird. They discovered they shared a common interest in automobiles, soon became friends, and the man took Bird with him on visits to Blackpool and Brighton. He even allowed Bird to drive his Riley, until he began to make extortionate demands.[52] The same year *The Times* reported two nineteen-year-old Lambeth laborers had threatened to make accusations against a wealthy man who took young men out in his automobile. Mr. A protested that he was only giving driving lessons.[53] In 1938, a laborer demanded five pounds from Mr. X. Mr. X had given him a ride in his motorcar, and again the question was if "improper behavior" resulted. When charged, the accused, who had a police record, asserted, "His type ought to be made to pay."[54] Obviously such mercenary relationships were not "typical" same-sex encounters, but their reportage confirmed the newspaper-reading public's understanding that one of the attractions of automobile ownership was that it drew to one the admiring attention of either sex.

The blackmail cases also made it clear that, in middle-class eyes, the sight of a laborer in an expensive car did not look right. In a class-ridden society, some men were not supposed to drive motorcars. In likening modern civilization gone awry to the image of a high-powered car with a monkey at the wheel, H. G. Wells provided a variant on this theme.[55] Likewise Lord Horder, the royal physician, pessimistically asserted that it was a sign that something was wrong in the world when "in Central Africa the Negroes also ride in motorcars and dose themselves with aspirin."[56] The British elite's obsession with the motorcar as a status symbol was understandable. In 1939, only one-fifth of the British population had a car.[57] Motorcars were first and foremost a symbol of wealth, and the idea that they should in effect be monopolized by the upper classes demonstrated their sense of entitlement. Speed was to be restricted to the elite. In his memoirs, the writer Douglas Goldring recalled how the Bright Young People flaunted such power. He claimed that their vices included driving "their cars about at high speed, when under the influence of drink, in the hope, if there was a smash, that the case would be reported in the Sunday newspapers."[58] Such crass materialism did not go uncriticized. George Sturt, a defender of rural crafts, wrote a friend in 1911, "I don't like England as she is—industrial, over-capitalized, where the Struggle to Live is so sordid, and success means motorcars and insolence."[59] C. E. M. Joad attacked motorists as the "most hated of all" violators of the English countryside, polluting while pursuing the undiscovered.[60] The view was echoed by E. M. Forster in *Howards End* (1910), where the Wilcoxes' motorcar desecrates the calm of village life.[61]

And, since wealth bought cars, in a culture of consumption, so too it indirectly bought sex. In H. G. Wells's *In the Days of the Comet* (1906), the car served as a symbol of sex and class power. The lower-class hero's courting of Nettie is jeopardized by his upper-class rival. "I was still meditating a tremendous retort, an indictment of all Lord Redcar's class, their manners, morals, economic and political crimes, when my trouble with Nettie arose to swamp all minor troubles. Yet, not so completely but that I snarled aloud when his lordship's motorcar whizzed by me, as I went about upon my long meandering quest for a weapon."[62] Music hall artistes understood such anger and envy. By flaunting their success through the purchase of expensive automobiles, the stars allowed their working-class fans in some sense to share in the delights of car ownership.[63] As Andrew Horrall notes: "Though audience members might have been too poor to purchase a car they enjoyed the status of ownership vicariously through the depiction of luxury vehicles. These aspirations were summed up by the title of the 1904 song, 'If you can't buy a motorcar, buy a motor cap.'"[64]

The British were awed by the availability of cars in America. In Britain, it was only in the 1930s that motorcars and highways were made accessible to the masses. In the mid-1930s, the average London house cost 500 pounds and a Rolls Royce 2,500 pounds. Yet an Austin Seven was only 165 pounds, and the Morris Eight and Ford were now also relatively cheap.[65] Conservative commentators denigrated the arrival of such inexpensive models as yet another sign of their society's decline. In *Vile Bodies,* Nina is made sick at sight of suburbs, "arterial roads dotted with little cars" empty factories, disused canals, wireless masts, and electricity pylons.[66] Fiction writers made clear the common notion that the type of car driven designated one's status. In *Brighton Rock,* the mob boss Colleoni has a sleek Lancia, whereas the young punk Pinkie drives "an ancient Morris . . . with flapping hood, bent fender and discoloured windscreen."[67]

Due to the cost of cars, most interwar British drivers were middle-class men. Because of the greater accessibility to automobiles in the United States, early twentieth-century American masculinity was refigured to include notions of mechanical ability. "Mechanical ability was becoming an attribute of gender, not social class," states Clay McShane. "The car became a masculine status object, driving a masculine skill, one of the few skills that fathers passed directly to their sons."[68] In contrast, in Britain where cars were long restricted to the elite, the chauffeur emerged as a curiously ambiguous figure—a lower-class male who in social terms was an inferior yet was more knowledgeable than his master. Who then was the more masculine?[69] In *Howards End,* Charles Wilcox is in constant battle with his driver Crane. In George Bernard Shaw's *Man and Superman* (1903), the "New Man" Tanner protests that he does not know how to deal with either his car or his cool, superior chauffeur. "I am a slave of that car and of you too. I dream of the accursed thing at night."[70]

More potentially worrying were the relations between the chauffeur and the lady of the house. Space was gendered. The middle-class woman's realm was the private, and the man's the public. With the arrival of the car, the question was immediately posed, was a woman taken for a drive in a public or private space? And if a male employee was driving, who was in control—the woman sitting at the back or the man at the wheel? Should a respectable woman be in a private space with a social inferior? The figure of the chauffeur crystallized such worries about how technological advances could erode traditional class and gender relationships. In *Howards End,* Margaret cannot escape her driver Crane though she regards him as impertinent and disloyal.[71] H. G. Wells begins *The Wife of Sir Isaac Harman* (1914) with the chauffeur haughtily insisting that he not be taken for a footman, and he goes on to tell the intimidated Lady Harman who is in charge. It is he who will decide how much time they have and how fast he will go.[72] A more comedic portrayal was offered in Stella Gibbons's *Nightingale Wood* (1938). Here the chauffeur not only runs off with one of the daughters, he is also left money by an employer leading some to think he was "the old man's boy friend." His wife says such rumors are understandable as the chauffeur is so good looking.[73] The chauffeur thus emerged as a new male sex symbol. Some fictional accounts attempted to finesse such threats. In Tracy Louis's *Sylvia's Chauffeur* (1911), the young woman, infatuated by the man she assumes is a chauffeur, is relieved to discover that he is in fact young Lord Medenham.[74] But the fear that affairs involving chauffeurs and mistresses were almost inevitable and would necessarily have disastrous results appeared to be validated by the famous Rattenbury case of 1935. The eighteen-year-old driver and the thirty-nine-year-old wife of a well-known architect (thirty years his wife's senior) began an adulterous relationship that led to the boy killing the husband.[75]

Given that so many commentators portrayed the motorcar as in effect an extension of a man's masculinity, what did they make of the woman driver? A point of entry into such questions is offered by Berta Ruck's popular novel *Sir or Madame?* (1923). It begins with some locals gossiping about a lady being wooed by her chauffeur and a widow who makes advances on an attractive young driver. But Ruck gives the chauffeur-as-womanizer theme a twist because the young man in her novel in reality is a high-spirited, cross-dressing young woman. Guelda, the daughter of a peer, had driven during the war, and in coat and breeches now passes as a presentable boy. "Smith" is taken on by young Sir Ralph who is impressed (as are the local girls) by the "dash" of the boy obvious in "Smith's" vitality, health, and adventurousness. The troubling specter of same-sex attractions hovers briefly but is quickly squelched when the masquerade is revealed. The potboiler ends with the predictable happy heterosexual coupling of master and would-be servant.[76]

Ruck's heroine was not the only fictional character to have gained driving experience during the war. Much of the plot of Radclyffe Hall's classic lesbian novel *The Well of Loneliness* (1928) concerns the chief character speeding across France in a light Ford ambulance. Stephen Gordon's first passion is for horses. As a girl, she daringly rides astride, and the master of the local hunt grudgingly admits she has a "fine seat." Her father buys the first family car, a Panhard, and employs a chauffeur who teaches her to drive. At the wheel, she again demonstrates her skill and courage. When she comes into money, she assumes further male prerogatives. She goes up to London, has her suits made by her father's tailor, and purchases pajamas and a man's dressing gown. Her buying a powerful automobile—the obvious acquisition of a wealthy young man about town—is her ultimate violation of stereotypical gender norms: "She ordered herself a rakish red car; a long-bodied sixty-horse-power Métallurgique. It was one of the fastest cars of the year, and it certainly cost her a great deal of money." When the war breaks out, she joins the London Ambulance Column and in France meets the love of her life, Mary, another driver.[77] After the war, Stephen buys herself a large "touring Renault" and "a smart little landaulette for Mary."[78] The roles played by the members of this same-sex couple are thus complemented by the cars they drive.[79]

The idea that a woman who drove might harbor same-sex desires was so current that Evelyn Waugh in *Vile Bodies* (1930) exploited for comedic purposes the image of a lesbian using her car as a means of seduction. Charity, the member of a gospel choir touring England, goes out for a drive in Mrs. Panrast's Rolls Royce. When asked why, Charity replies,

> "Well, if you must know, I thought she *was* a man."
>
> "Thought she was a *man,* Chastity? That doesn't sound right to me."
>
> "Well, she looks like a man and—and she *goes* on like a man. I saw her sitting at a table in a tea-shop. She hadn't got a hat on, and I couldn't see her skirt . . . ooh . . . how can I tell you if you keep pinching . . . and she smiled and so, well, I went and had some tea with her, and she said would I go out with her in her motorcar, and I said yes and, ooh, I wish I hadn't now."[80]

In *Orlando: A Biography* (1928), Virginia Woolf naturally presents her androgynous leading character, now living in the twentieth century, as an expert at the wheel who drives "masterfully, swiftly."[81] The novelist playfully suggested that her own Singer motorcar was sexually indeterminate. She wrote her sister Vanessa, "I can't believe your amazing stories of the Male and Female parts of the Renault. Do the French sexualise their engines? The Singer I know for a fact to be hermaphrodite, like the poet Cowper."[82]

Whether they were hostile or supportive, commentators concurred that the automobile could play a role in expanding woman's sexual freedoms, yet in 1933 only 12 percent of driving licenses in Britain were held by women.[83] Nevertheless a beautiful young woman at the wheel of a powerful car became an iconic image of modernity. Tamara de Lempicka's 1925 self-portrait in her green Bugatti immediately comes to mind. Or the death by strangulation of the American dancer Isadora Duncan in 1927 in Nice when her scarf became entangled in the wheels of an Italian roadster.[84] Those who presented images of sophisticated women seeking to subdue sullen machines played up the beauty and the beast motif.[85] Richard Hannay, the swashbuckling hero of John Buchan's *Greenmantle* (1916), has no trouble subduing sadistic Germans and Turks, and yet is cowed when driven off by the beautiful Hilda von Einem: "I had never been in a motor car with a lady before, and I felt like a fish on a dry sandbank."[86]

Driving may have reinvigorated masculinity, but many observers worried about the linkage of cars, conspicuous consumption, and the sexual liberation of women. A woman in command of a fast car conjured up notions of the sort of female recklessness or rebellion that Europeans first saw advancing in America.[87] Georges Duhamel, in *Scènes de la vie future* (1930), translated as *America the Menace* (1930), was impressed on the one hand by the painted and powdered American girls who began their "puberté automobile" at sixteen and on the other by the blighted countryside full of wrecker yards. For Duhamel, the modern goddess was epitomized by the young American woman who drove him home one night in a powerful motorcar, her skirt riding up her thigh, one hand on the wheel, the other holding her cigarette.[88] Only in modern times could this combination of mechanical power, female independence, and sexual accessibility provide such an erotic charge. Similarly in Evelyn Waugh's *Black Mischief* (1932), one of the proofs offered of the young Basil Seal's fascinating personality is that as an Oxford undergraduate, "Lovely young women visited him from London in high-powered cars."[89]

The car—promising independence, escape, and an enlargement of the private sphere—was a key symbol of woman's emancipation. In a 1931 speech, Virginia Woolf claimed that it was her desire for a motorcar that led her to becoming a writer. "I made one pound ten and six by my first review; and I bought a Persian cat with the proceeds. Then I grew ambitious. A Persian cat is all very well, I said; but a Persian cat is not enough. I must have a motorcar. And it was thus that I became a novelist—for it is a very strange thing that people will give you a motorcar if you will tell them a story."[90] She was overjoyed by the secondhand Singer that she and her husband Leonard acquired in 1927. In her diary, she at first expressed caution: "It will I think abolish loneliness, & may of course imperil complete privacy."[91] But she was quickly seduced. "Yes, the motorcar is turning out the joy of our lives, an additional life, free & mobile & airy. . . . Soon we shall look back at our pre-motor days as we do now at our days in the caves."[92]

In playing off the image of the woman as "Angel in the House" against that of woman as car owner, Woolf was one of many interwar commentators who wondered if the traditional family could withstand the corrosive forces of mobility unleashed by the automobile. Harry Joseph Birnstingl, in *Lares et Penates; or, The Home of the Future* (1928), claimed that the motorcar was the "embodiment of that unrest and desire for continual movement and excitement. Indeed the motorcar has become such a necessity that the house is sometimes nothing more than an annex to the garage."[93] The architect John Gloag concurred that the fixation families in the 1930s had on the car were culturally dangerously. "The mobility that motoring has brought to thousands of people tends to minimize their criticism, and even to obliterate their consciousness, of restricted space in their homes."[94] The growing popularity of the "small service flat," declared Gloag, was a symptom of the decline in importance of the middle-class family home and garden.

Interestingly enough the novels of the interwar period that most vividly portrayed the relationships of woman and cars ended with the drivers' deaths. Michael Arlen's *The Green Hat* (1924) is replete with anthropomorphic references to cars. He describes a "clever little Citroën taxi" and "the burning eyes of the Renault."[95] Iris Storm, his melodramatic heroine, first appears in a Hispano-Suiza. "This car charmed the eye. Like a huge yellow insect that dropped to earth from a butterfly civilization, this car, gallant and suave, rested in the lonely silence of the Shepherd's Market night."[96] Her young husband having committed suicide, Iris now attempts to lose herself in cigarettes, affairs, and fast cars. "I am not the proud adventuress who touches men for pleasure, the silly lady who misbehaves for fun. I am the meanest of all, she who destroys her body because she must, she who hates the thing she is, she who loathes the thing she does"[97] And indeed she ends her life driving her "great beast" of a car directly into a tree.[98]

The Green Hat was a best seller. Norah C. James's *Sleeveless Errand* (1929) was seized by the police and declared by a magistrate to be an obscene publication.[99] In point of fact, the two novels were cut from much the same cloth. In James's book, Paula is the world-weary beauty now jaded by drunks, whores, absinthe drinkers, and other decadents. She complains that her current circle of friends are either promiscuous or "homos." A companion concurs: "What's wrong with us? All sex-mad—sex, sex, sex."[100] Paula describes herself as a member of the damned generation who have lost their possible marriage partners in the war. Their immorality is now aped by their younger sisters. "We sneer at goodness. . . . We're bored with people who aren't bawdy. We call them pigs and prudes if they don't want to talk about copulation at lunchtime and buggery at dinner."[101] She suggests to Bill that as a suicide pact they will drive off a Sussex cliff and hires a large Vauxhall. At the last moment, she has a pang of conscience, convinces Bill to return to his wife, and takes her last drive. The novel ends with the line: "The car rushed forward into space and began to spin over in the air as it fell."[102]

Even a serious novelist like Elizabeth Bowen followed this script. The quintessentially independent female character in *To the North* (1932)—single, partner in a travel agency, and owner of a small car—orchestrates a head-on collision in which both she and the cad who seduced her are killed.[103] These cautionary tales, in implying that mobility undermined female morals, linked the motorcar and female sexuality. They served as a reminder that society took a man's interest in speed to be natural, but to describe a woman as "fast" was to imply that she was promiscuous or at the least sexually active.[104] Such novels denigrated changing moral standards and, though providing titillating accounts of sordid deviant behavior, ended with expressions of concern for the propping up of home and family. Of course, it is a moot point as to how impressed the readership was by such carefully crafted conclusions.

The image of the motorcar, particularly as depicted by fiction writers, thus served a variety of functions. It could represent male power and female independence, the assurance of upper-class entitlements and the threat of lower-class mobility. To the extent that its advent was turned to the purposes of explaining perceived changes in virility, courtship, seduction, and divorce, the car played a central role in reconfiguring twentieth-century sexuality.

Let us now shift gears and note that while it is well established that interwar commentators described cars in anthropomorphic terms, that they were sexualized, the other side of the argument—that in an automobile age people increasingly viewed themselves as machines—has been less observed.[105] That is to say, the same culture that made motorcars sexy, made sex "automobilic."[106] It was not surprising that doctors, sexologists, and laypersons came to cite the car as a cause of a whole range of sexual problems. More problematical was the fact that many turned to automobile metaphors in seeking to explain sexual performance.

In popular literature, the ability to own and drive a car was, as we have seen, commonly portrayed as a sign of potency. The counterargument, that the automobile could threaten virility, was also made. Freud held that attempts to repress the sexual feelings associated with traveling could lead in adult life to "railway anxiety." Excessive stimulation might lead to a neurosis, manifested in a disorder "in the sexual mechanism or chemistry."[107] The American sex reformer William Robinson warned, "Fast automobile riding has recently been brought forward as a cause of impotence, and there are good grounds for believing that this is a real, and not a fanciful, cause. The jarring, the constant worry and anxiety which are inseparable from fast and furious automobile driving, induce a state of neurasthenia that is responsible for impotence. The impotence is generally of a temporary character and yields to treatment at once, when the sport, or rather the furious speed, is given up."[108] A hostile critic of the car, Victor Vecki was obviously pleased to report that the vehicle created male sexual dysfunctions. "The long suffering public, pedestrians and horse-back riders, who can never tell when a drunk and speed-crazy hoodlum in an automobile is going to crush the life out of them, may feel some satisfaction by knowing that motorcar speeding causes sexual impotence." While some believed that the mental concentration required in speeding caused neurasthenia, Vecki held that "the jar of the car and the bouncing upon the soft and warm upholstery are the chief harmful influences, because the speeders that I have so far observed certainly had no mind worth to speak of to concentrate on anything."[109] In her popular marriage manual, *Enduring Passion* (1928), Marie Stopes spread the word that motoring could indeed threaten virility.[110]

Potency was not the only qualification required to marry. Given that car ownership was policed, some asked why marriage was not similarly governed. Sex experts argued that people, like cars, needed inspections. "It is most desirable," argued Gladys Cox, "that there should be a medical overhaul before marriage, not only to ensure a clean bill of health, but also to detect any physical abnormalities that would be an obstacle to the normal sex act and would therefore prevent the proper consummation of the marriage."[111] Claud Mullins, a police court magistrate, complained that it was easier to get a wedding license than a driver's license. "When we apply for licenses to drive a motorcar we receive nowadays a booklet on how to drive. The booklet is furnished, I believe, by voluntary agencies interested in the prevention of accidents. It should be possible to compile a little book on Conduct in Marriage and Parenthood, which the Registrar could give to all applicants for marriage at the time that notice of the intended marriage is handed in."[112]

A widespread complaint was that once married, men made it clear that they knew more about their cars than about their own bodies.[113] According to marriage counselor David R. Mace, "If husbands would seek to understand the technique of love-making as thoroughly as they understand the mechanism of their cars, and wives took as much interest in the art of being responsive lovers as they do in that of being fashionably dressed, we would find ourselves living in a very much happier world."[114] In her futurist novel *Man's World* (1926), Charlotte Haldane had a character complain that her husband "had been educated to be interested in the inside of a motorcar, but his own inside he dared not think about."[115] Haldane envisaged such ignorance being banished in the forthcoming society where youths would have access to sex education, geneticists would carry out embryonic research, and the government would set demographic targets. An American writer agreed that most men took the male sex function for granted. When impotence hit, they turned to doctors and drugs, just as they went to their auto mechanic to repair their motorcar.[116]

Sex experts repeatedly employed automotive similes. Once in the marriage bed, Margaret Sanger's advice was that the husband should "like a skillful driver, at every moment hold himself under intelligent control." This argument was countered by a contributor to the medical journal the *Lancet*: "For a man to attempt restraint," he warned, "is equivalent to trying to drive his car at full speed with the brakes on. Men, like motorcars, do not last long under such treatment."[117] "Congestion" was as much a threat to smooth flowing traffic as it was to a woman's sexual response. Isabel Hutton held that it was up to the bridegroom to gently woo his mate. "If the woman has no orgasm, then her sexual organs are left in a state of congestion which may take many hours to pass off."[118] And although the sex expert Kenneth Walker stated that love was an art, not a science, he described it much like a road race. The man was inherently faster, and it was his task to slow down. It took time for the woman "to attain erotic efficiency." Walker concluded optimistically: "With practice, however, she will learn to accelerate her orgasm so as to reach it at about the same time as her partner."[119] Marriage counselors presented themselves as qualified to provide the public with such information on the mechanics of sexuality. If your car stalls, wrote Mace, you cannot repair it if you don't know how it functions. So too, couples could not fix their sex difficulties if they did not understand how they had arisen.[120]

It has been argued that we Westerners like "to define our humanity against machines," but one is struck by how many interwar writers appeared to hope that people would become more like cars.[121] "No man treats a car as foolishly as he treats another human being," wrote Bertrand Russell. "When the car will not go, he does not attribute its annoying behaviour to sin; he does not say 'you are a wicked motorcar and I shall not give you any more petroleum until you go.' He attempts to find out what is wrong and to set it right."[122] In a 1922 article, Julian Huxley hopefully predicted that the hospitals of the future would be like car garages or "human repair shops": "Men will have found methods for keeping organs alive outside the body, or they will be able to make grafts from tissue-cultures. Thyroids, pituitaries, adrenals, pineals, interstitial tissue, and many other regulating organs now unknown, will be in their several places, and aging humanity will come in to have their bodily system reanimated as cars come in to a garage to be overhauled."[123]

With the couple married and medically serviced, the expectation was that children would soon result. Failing to reproduce was blamed on the car. Such materialistic concerns, asserted John Galsworthy, led the Forsytes to opt for small families. "Sooner in fact than own children, they preferred to concentrate on the ownership of themselves, conforming to the growing tendency—*fin de siècle,* as it was called. In this way, little risk was run, and one would be able to have a motorcar."[124] Walter M. Gallachan agreed. "Love is so undervalued as a source of happiness, a means of grace, and a completion of being, that many men would sooner work to keep a motorcar than to marry."[125] In the Depression of the 1930s, the British fertility rate plunged to an all-time low. Many believed that it was not simply a coincidence that family size declined as car ownership increased. In the United States, the familiar joke had it that the American family had shrunk to fit Ford's four-seat Model T. In Britain, a somewhat similar jest was made; one had to decide between having a baby and buying a Baby Austin.[126] Norman Long sang of "My Little Austin Seven" (1932):

> It matters not at all you think it's awful small
> Believe me boy, it's in
> If you want to give the wife a ride without all her relations
> You'll never beat an Austin Seven

Those who favored family planning employed automobile metaphors in a positive sense.[127] People once had no ways of preventing pregnancies, but, according to Marie Stopes, birth control changed all that. "In the control of conception the human race has an instrument comparable with the steering wheel of a motorcar, able to direct progress to the right or the left or straight on."[128] Constructive birth control, in allowing the young and fertile to space conceptions, would aid the race and the state.

But fears were expressed by some in the 1930s that the birth controllers had been all too successful and Britain faced depopulation. The demographer Enid Charles likened the nation's decline in fertility to "a sports car . . . rushing downhill without brakes."[129] In a poem that the independent member of Parliament A. P. Herbert wrote to greet the 1937 Population (Statistics) Bill, the money spent on automobiles was blamed in part for Britain's demographic slump:

> Five hundred brand new motorcars each morning rode the roads
> And flashed about like comets or sat as motionless as toads;
> Whichever course they took they made the public highway hell
> *And everybody wondered why the population fell.*[130]

Alice Jenkins, an abortion reformer, agreed that the huge expenditures on cars, the lack of public transport, and the worshipping of "mechanized speed"—despite the number of child deaths in road accidents—were all reasons why parenthood was avoided.[131] Though the Anglican Church in the 1930s came to accept the legitimacy of birth control, "the Mothers' Union advised that 'all young couples should be willing to have at least one or two children to complete their marriage' and urged members to 'strive to restore a public opinion which will rate the possession of children higher than owning a motorcar or a comfortably furnished house.'"[132] The eugenicist Charles Galton Darwin agreed that "the prospect of owning a motorcar is a sufficient bribe to sterilize most people."[133] A postwar sociologist researching marriage reached a similar conclusion. He asked his survey subjects, "What is better, a car or a wife? 'I was courting,' confided one, 'and should have been married by now, but instead I bought a car.' At first I treated this as a joke, but I heard it so many times in so many versions ('I can't afford a car and a wife, so I drifted away from my girl') that I had to regard it as a major issue for youth at present."[134] Such accounts were in effect arguing that young people were employing the car as a sort of birth control device.

Brian Stableford has argued that the science-fiction writer S. Fowler Wright was unusual inasmuch as "the sickness of civilisation was symbolised for him by the motorcar, the idea of birth-control and the pursuit of comforts. The motorcar stood for the march of science and the mechanisation of the human world, birth-control for an attitude of mind by which men set themselves in opposition to nature, and the pursuit of comforts for a smokescreen which blinded men to the futility of idleness."[135] But this pessimistic linking of sex and the motorcar was not uncommon.[136] In D. H. Lawrence's *Lady Chatterley's Lover* (1928), Mellors condemns mechanization for emotionally stunting the masses. "Their spunk's gone dead—motorcars and cinemas and aeroplanes suck the last bite out of them."[137] And in writing of tourists in Italy, Lawrence scoffed,

> But the blatant bathers don't know, they know nothing;
> the vibration of the motorcar has bruised their insensitive bottoms
> onto rubber-like deadness, Dunlop inflated unconcern.[138]

Drawing on Freud, Henry Durant, a social investigator, argued that as labor became less satisfying the attention of the masses' libido shifted "from work to sex."[139] Spontaneous popular leisure time activities were usurped by what Durant bleakly called the "machinery of amusement," meaning drink, cinema, football, racing, and gambling. He included the fixation on the car as part of this trivialization of culture. Wilfred Wellock agreed that in a machine society the goal was to eliminate the worker's intelligence. "Accordingly the spiritually restless, soul-starved worker, lured by emotionalism and speed, is rarely happy unless he is wallowing in sexual or other excitements or racing over the earth in a high-speed car."[140]

Other progressive writers bemoaned the masses' fixation on cars and sex. Orwell noted that by the 1930s the leading contemporary writers—Joyce, Eliot, Pound, Huxley, Lawrence, Wyndham Lewis—no longer believed in progress. "They don't any longer believe that men are getting better and better by having lower mortality rates, more effective birth control, better plumbing, more aeroplanes and faster motorcars."[141] Winifred Holtby lamented the fact that "the present generation demonstrates an enormous interest in sex"—manifested in psychoanalysis, dislike of virgins, and racy novels—and opined that "if one sort of revival is more probable than another in the near future, it is a revival openly and avowedly preaching sexual salvation." Her view was that a life dominated by the libido was as artificial as one dominated by the motorcar. "Man cannot live by bread alone; but neither does he live like an internal-combustion engine by a perpetual series of small sexual explosions."[142]

Many took the motorcar and the contraceptive as the two key icons of modernity. It was, therefore, hardly surprising that so many interwar commentators regarded automobility and sexuality as necessarily linked. Talking about cars licensed writers to address a variety of sexual issues that the prudish preferred to skirt. Though cars obviously had an enormous economic and social impact, they more importantly spoke to the emotional desires of many. One, of course, needs to be cautious in asserting exactly what psychological function they served. Jonathan Dollimore has warned, "The point is that nothing invites third-rate sex-psychology so much as cars. . . . The car is just one of many 'desirable' objects which disappoint desire even as they ensnare and incite it; not the end of dissatisfaction, but its vehicle."[143] For some men and women, it promised liberation and escape. For others, the motorcar represented all that was wrong with a consumerist, mechanized society. And for still others it functioned with an efficiency that humans could only envy. If the representations, images, and narratives of automobility varied enormously, those who crafted them tended to agree that, for better or for worse, the motorcar was playing a key a role in reconfiguring sex, gender, and reproduction. The only machine that rivaled it in cultural significance was the robot.

CHAPTER 3

"A race of human machines"

Robots and Reproduction

In 1923, the London literary scene saw the appearance of two innovative works. In April, the play *R.U.R.; or, Rossum's Universal Robots* by the Czech dramatist Karel Čapek was presented at St. Martin's Theater; in May Heinemann brought out E. V. Odle's comic novel *The Clockwork Man*. Odle's entertaining book is now all but forgotten whereas Čapek's three-act drama is still in print. Thanks to it, the term "robot" entered the English language, and such humanoids were immediately hailed as icons of modernity.[1] When in 1930 the centenary of the railway was celebrated, Liverpool's Lime Street station was surmounted, not by a locomotive engine, but by "a robot figure symbolizing the Age of Machinery."[2] The centerpiece of the 1932 Radio Exhibition at Olympia was a robot of chromium-plated steel called "Alpha," which, it was claimed, could read a newspaper and tell the time.[3] In the interwar period, a host of commentators wrestled with the challenges posed by such "mechanical slaves." Why such a fascination? The most obvious answer is that the robot represented the displacement of humans by machines. In what follows, a sense will be given of the extent to which those worried about the coming of a mechanized society batten on the robot. But the main thrust of this chapter is that fears of mechanized men were driven by more than concerns about technology. In looking closely at the cultural context of the interwar years, one finds that such discussions were heavily colored by eugenics. It took little imagination for those alarmed by reports of the fertility decline of the middle classes to see the robot as representing both the docile worker on which modern industry increasingly depended and the unthinking drone whose uncontrolled reproduction posed a serious threat to social stability.

Let's begin with the play.[4] *R.U.R.* takes place in a factory on an island sometime in the future. We are told that old Rossum discovered a way to create living matter, something like protoplasm in a test tube, and vainly tried to make actual men. His nephew young Rossum, an engineer, restricts himself to producing simple working machines—which boast no ornaments and have no needs—the Robots.[5] Put together like motorcars, their livers and brains made up of paste, their nerves and veins spun, these beings are produced biochemically. They are not electrically or mechanically driven. The handful of men who oversee their production is joined in act 1 by a naive young woman, Helena Glory, who expresses her horror at the cruel exploitation of the robots. The men scoff at her notion of liberating them and explain to her that robots feel no pain, have no soul or desires. Moreover, being produced by the thousands, they will soon do all the world's work. In the resulting utopian leisure society, humans will finally be able to perfect themselves. Hovering in the background, however, are both the specters of overproduction and the robots' inexplicable restlessness, manifested by the gnashing of teeth and their cramping up.

Act 2 takes place five years later. Helena, who has married the factory manager Domaine, hears that in Europe, robots, used for destructive military purposes, have suddenly revolted against their human masters.[6] In France, the insurgents issue a "manifesto" calling for all the robots of the world to rise. A world revolution ensues, which the robots, enjoying a numerical advantage over humans, easily win and then turn their murderous attentions to Rossum's factory. But as the robots are sterile, have a life span of only twenty years, and do not possess the secret of their manufacture, the managers believe that they still have a vital bargaining chip. They are accordingly horrified to hear that Helena has destroyed the formula. They also learn too late that the cause of the revolt is that Dr. Gall, the factory physiologist, gave the latest generation of robots the ability to feel pain. In making them more like people, he gave them the rudiments of a soul.

In the last act, thousands of robots lay siege to the factory. All the humans are killed save for a craftsman whom the robots almost consider one of them, as he works with his hands. They desperately ask him how they might reproduce. Though he cannot help them, he is surprised to discover a male and a female robot who demonstrate their love for each other. The play ends with him sending this new Adam and Eve out into world.[7]

Though critics judged the play to be uneven, it enjoyed an enormous success in London.[8] Toy robots were soon on sale. Robots figured in various skits and jests. The tabloid press ran cartoons of them subjected to pointless pursuits such as playing golf and making beds.[9] Advertisers exploited the craze. "A Robot is the only type of man who doesn't enjoy Schweppes," declared the soda water manufacturer.[10] The moral of *R.U.R.* was debated by G. K. Chesterton who took the play to be a salutary warning against overindustrialization and George Bernard Shaw who held that, on the contrary, if one wanted more leisure, as he did, it was necessary to "Robotize" the world.[11]

The word "robot" suddenly crystallized concerns about the relationship of humans and machines. Overnight robots were discovered to be everywhere. Such was the newspapers' fascination with the subject that they now dubbed almost every labor-saving device a "robot." Sir Leo Chiozza Money, who shared Shaw's enthusiasm for the machine, was reported as praising "tea robots" that did the boring work of filling packages at the Lyons factory.[12] According to the *Daily Mirror,* the Westinghouse Electrical Company used "mechanical men" to turn on lights.[13] The slot machines that dispensed food, radios that offered five minutes of wireless service for a penny, money-changing machines, telephone message recorders, automatic gearshifts, and the changing traffic lights on Oxford Street were all now called "robots." The public learned that there were robot bookmakers, robot self-massaging suction rubber roller devices used to lose weight, robot wireless operators on ships, and robot pilots (or gyroscopes) for long distance flights.[14] Robot mechanical harvesters replaced laborers, and robot accountancy machines displaced clerks. In 1930, the *Daily Mirror* chided the Royal Academy that if it were to be up to date in an age of speed and invention, it should have put on display a portrait of a robot.[15] And if foreign sports stars like René Lacoste were not robots, the British press gave them the backhanded compliment that they played with the same cold, mechanical efficiency.

Epitomizing as they did the unthinking and disciplined, robots could be easily imagined serving the forces of order. The novelist Richard Aldington referred to the World War One soldier as "merely a unit, a murder robot, a wisp of cannon fodder."[16] Major-General J. F. C. Fuller ended his book on the future of warfare by lamenting the fact that when soldiers were eventually replaced by robots that felt no fear and thus could manifest no heroism the glory of battle would be gone forever.[17] Cartoonists portrayed robots acting as policemen and civil servants.[18] A Canadian author extrapolated from the fact that the Germans employed a registration system known as the Meldewesen, which required everyone entering a town to report to police, to spin out the tale of a "robot detective." In the novel, a retired policeman feeds facts into a machine modeled on "the great man-hunting

Robot of Berlin."[19] In a grating voice, it demands the pertinent information that finally results in the solving of a series of drug-related murders.

The press referred to the talkies—which displaced silent films—as an example of the application of "robotism" to art. A critic asserted that to be moved by a film with a sound track was like falling in love with a robot. Unlike live theater, celluloid had no soul.[20] The theater was fascinated and repelled by the machine.[21] In 1930, a ballet called "Fleet Street" with dancers dressed as robots was put on by the Intimate Review at the Duchess Theater, but the self-proclaimed defenders of the arts naturally enough lamented what Lord Cavendish-Bentinck described as "a mechanical and Robot age."[22] D. H. Lawrence had frequent recourse to the metaphor, attacking "Robot Feelings" and "robot democracy."[23] In the poem "The Gods! The Gods!," he envisages the beach covered with "great robot limbs, robot breasts, robot voices, robot even the gay umbrellas." In "Sea-bathers," he conjures up a vision of bodies made of "india-rubber tubing inflated / and the half-hidden private parts just a little brass tap, rubinetto, / turned on for different purposes."[24] He went on to present contemporary society in *Lady Chatterley's Lover* (1928) as dominated by the machine: "Each man a machine-part, and the driving power of the machine, hate." Clifford Chatterley, tied as he is to a wheelchair, is portrayed as a sort of android and the miners as in effect robots violating mother earth.[25]

Conservative defenders of the arts lamented the fact that some artists seemed to be embracing the machine. "The Cubists appear to have been those most affected by the mechanical world about them," moaned Professor Henry Tonks of the Slade School of Art. "If they expressed men at all, they seemed as if they were anticipating the Robot; sometimes in the composition it was difficult to find any sign of life."[26] The painter and author Wyndham Lewis sneered in "Shropshire Lads or Robots Again" that Marxism and machine-mindedness went together in young British poets like Auden and Day-Lewis but that their adoration of industrialization—common to fascist or communist states—should have been out of date in twentieth-century Britain.[27] The American writer Edward J. O'Brien made the point that writers could not escape mechanization. "The machine also taught them to think of sex in purely mechanistic terms. For the satisfaction of the sex instinct, men were regarded as interchangeable parts, and so were women. . . . The sex act was even pictured as a mechanical function, and the current language of obscenity used machine images constantly. The modern heaven, in T. S. Eliot's sardonic phrase, was a 'promise of pneumatic bliss.'"[28] Given that the fantasist J. R. R. Tolkien sought inspiration in the Middle Ages, it was hardly surprising that he too held machine culture in contempt. "The maddest castle that ever came out of a giant's bag in a wild Gaelic story," he asserted, "is not only much less ugly than a robot-factory, it is also (to use a very modern phrase) 'in a very real sense' a great deal more real."[29]

The word "robot" also entered the political vocabulary. Lord Henry Cecil (Oxford University, Unionist) opposed the extension of the franchise in 1928 as "Mechanizing Parliament." A mass electorate was not democratic, he informed the House of Commons, "It was government by bosses of an elective assembly of robots. (Laughter.) Those robots were qualified in proportion to their efficiency for various honours at the disposal of the Crown, and some of them who were very good robots went to the other House of Parliament, there to continue the mechanical process. (Laughter.)"[30] A similar argument was made by Sir A. Daniel Hall, an agricultural expert: "Day by day we see how robots can be created by a Government intent only upon power, which can control the Press and all publications within its country, which can close the frontiers to other communications, which has learnt to utilize wireless and all the machinery of popular advertisement and above all education itself as instruments of propaganda to run people into a mould."[31]

This interwar fascination with robots is in some ways perplexing inasmuch as automatons were not new; they had been constructed and discussed for centuries past.[32] Playing with the notion that the state sought to instill passivity in its subjects, the radical poet Shelley wrote in the early 1800s: "The man / Of virtuous soul commands not, nor obeys: / Power, like a desolating pestilence, / Pollutes whate'er it touches, and obedience, / Bane of all genius, virtue, freedom, truth, / Makes slaves of men, and, of the human frame, / A mechanised automaton."[33] In the 1830s, Heinrich Heine, the German romantic poet, envisaged the English producing a mechanical man.

> The tale goes that an English inventor, who had already invented the most ingenious machines, finally hit on the idea of constructing a human being. In the end he succeeded; the work of his hands could behave and act just like a man; it even bore within its leathern breast a sort of human feeling differing not too greatly from the usual feelings of the English; it could communicate its emotions by articulate sounds, and it was precisely the noise of the wheels inside, of springs and screws, which was then audible, that lent these sounds a genuinely English pronunciation. In short, this automaton was a perfect gentleman, and nothing was missing to make it a real human being except a soul. This, however, the English inventor could not give him, and the poor creature, aware of its deficiency, tormented its creator day and night with the plea of giving it a soul.[34]

The moral of Heine's story was a mechanical existence did not suffice.

In the short story "The Sandman" (1815), E. T. A. Hoffman presented a Professor Spalanzani creating an automaton called Olympia whom he successfully passes off as his daughter.[35] Mary Shelley, whose *Frankenstein* (1817) was the most famous portrayal of a cyborg, similarly drew on eighteenth-century interests in the concept of the man-machine.[36] Peter Wollen makes the point, however, that Victor Frankenstein's monster, the Golem, and the Homunculus—individual, almost magical beings—were qualitatively different from the assembly line products such as the robot and reflected another of society's cultural concerns.[37] Early automata performed, terrified, or amused; the robot does not seduce or entertain, it labors.[38] Whereas fears once focused on the powers of witches or mad professors, modern worries were raised by technologies that allowed machines to ever more successfully impersonate humans.[39]

"Robot" was originally a Slavic term for forced labor. In so naming his automaton, Čapek underscored its chief function. In the play, Rossum's nephew focuses simply on producing an intelligent machine—"a worker with the minimum amount of requirements."[40] Unsurprisingly, the threat of the robot was seized upon in the interwar period by those hostile to the mechanization of society. *The Times* reviewer of *R.U.R.* informed his readers that the robots, like Frankenstein's monster, inevitably turned on their makers. The play's tragic ending, he suggested, was the inevitable result of a community falling under the sway of industrial psychology as currently preached and practiced in the United States.[41]

Would British workers be reduced to such a level? Some declared that it was not only inevitable, but necessary. In the mid-nineteenth century, Hermann von Helmholtz dreamed of the machine freeing humans. "We no longer seek to build machines which shall fulfill the thousand services required of *one* man, but desire on the contrary that a machine shall perform one service, and shall occupy in doing it the place of a thousand men."[42] At the beginning of the twentieth century, economists like Alfred Marshall looked forward to the day when even the humble would have "mechanical slaves" that would make life easier for all.[43]

Defenders of machines went on to argue that those who tended them would naturally have to work with a machinelike efficiency. Following the dictates of scientific management, as popularized by Frederick Taylor's 1911 study, the task of the overseer was to train and select the worker for the task best suited for him or her—not allow the workers to think for themselves.[44] It might be remembered that with so many men having been mutilated in the 1914–18 war and requiring prosthetic limbs, new hybrid beings of flesh and aluminum were created and entered the British labor market. Amputees were shifted to appropriate tasks and their functions, if not their limbs, were restored. They demonstrated the truth of a central tenet of Taylorism, that industry required not entire employees but only "hands."[45]

Industrial psychologists like Bernard Muscio actually argued that the less thinking the worker did the better. "We may say, generally, that methods of work which substitute order and habit for judgments and decisions decrease fatigue, and in this respect alone, influence output beneficially." The manager's goal was to reduce both the workers' physical and mental energy, thus embracing "the principle of substituting automatisms for frequent acts of decision." Opponents might say the worker was thereby reduced to a mere "automaton." Muscio replied that such training did not necessarily lead to the "mechanising" of a man, but even if it did in some slight way, it was a fantasy to believe that any worker was ever free.[46]

A few months after the premier of *R.U.R.*, *The Times* ran a story titled "Fatigue in The Factory. The 'Fordising' of Industry. Welfare Workers' Views." It reported that a conference on industrial welfare at Balliol College, Oxford, had heard some oppose "Fordism" as soul destroying while others supported such modernization. "What we want is a Robot, and not a man," Miss Matthias, of the English Electric Company, was quoted as saying. "We do not want men of intelligence if we are going to Fordise industry. Let the workers work according to plan, and let them have a share in making that plan."[47] Henry Ford's assembly line means of production had by the 1920s become a byword for mindless, repetitive labor. An observer of the American automobile industry noted skilled workers were declining in number while the semi-skilled, who tended machines, increased. "The task of the worker requires simply speed, dexterity, alertness and nervous endurance to carry the 'robot' through dull, monotonous, fatiguing, relentlessly automatic operations."[48] Ford responded by insisting that machines did not destroy creativity; they provided the leisure time to allow it. He devoted a chapter to "Machinery: The New Messiah" in *My Philosophy of Industry*. Far from being embarrassed by the notion of treating humans like machines, he suggested in the section, "Repairing Men like Boilers," that just as rust-proofing prolonged engine life, so too could humans benefit from clean living.[49]

The poet Ronald Campbell Macfie, who contributed to Kegan Paul, Trench and Trübner's popular "To-Day and To-Morrow" futurist series, voiced the common concern that employers sought only the breeding of "morons and Robots with big muscles."[50] Responding to Macfie in the same series, H. Stafford Hatfield noted that the popularity of fictions like *Frankenstein* and films like *Metropolis* showed the public's fear of robots. The truth, however, was that unskilled laborers doomed to boring repetitive tasks were in effect already "semi-automata."[51] Advanced industry was experimenting with devices employing light sensitive cells, thermostats, and sound recognition components. It did not seek humanoid robots but dependable workers and efficient machines. Instead of trying to improve the working conditions of semi-skilled laborers, Hatfield argued—as did Ford—that humanitarians should support technologies that would replace them.

Factory hands were not the only workers who might be displaced. It says something of the class position of many of those in Britain involved in the discussion of machine culture that they repeatedly returned to the notion of replacing servants with appliances.[52] In *Antic Hay* (1923), Aldous Huxley presents a character who hates dealing with the lower classes. "Ah, if only they'd invent servants with internal combustion engines," said Bruin, almost pathetically. "However well trained they are, they always betray their humanity occasionally."[53] For some, this was not a joke. In a chapter titled "The Robot Age," A. M. Low promised his readers that the future world would be serviced, not by androids, but by machines devoted to the repetitive work such as that now carried out by nursemaids, chauffeurs, and gardeners. He felt obliged to reassure his readers, however, that, as such machines could not think, they would never rebel.[54]

The British attributed attempts to subject workers to behaviorist training and conditioning to make them mere adjuncts of the machine to "excessive Americanism," but in Bolshevik Russia there were similar enthusiasts. René Fülop-Miller reported in 1927 that such was the adulation of machinery in the Soviet Union that a historian seriously referred to Lenin as a "special appliance" while "a Bolshevist poet attempted to describe him as a 'greater screw' within the collectivist machine."[55] The factory poet Alexey Gastev insisted, "We must not only love the machine." He oversaw an Institute for the Scientific Organisation of Work and the Mechanisation of Man. Like Taylor, Gastev sought to establish laws of movement and stimuli to determine optimum work and rest periods. With their superfluous movements eliminated, workers were to be trained to be "labour machines" or "management regulators'" or part of the "directive apparatus." "This principle of organization is extended to all physiological elements, and thus a 'rhythmic rotation of work' is produced, which not only completely does away with all disturbing caprices and eccentricities of the nerves and soul, but removes all constitutional mental obstacles. The machine man is produced—and guaranteed to function properly."[56] The hope was that some day "instruments" (meaning workers) would, like electrical appliances, be artificially constructed.[57]

In Britain, many public commentators were stridently opposed to Fordism and Taylorism. The problem, as they saw it was not that machines were becoming more like humans, but that humans were being treated increasingly like machines. Such concerns spanned the political spectrum. On the right, R. Austin

Freeman (an anti-Semitic physician and popular detective-story writer) complained that mechanization now gave everyone access to "power." As a result of machinery, Britain was lumbered with the overproduction of shoddy goods and a homogenized, restless, unionized working class. With craftsmanship destroyed, the masses became no more than passive consumers. Racial deterioration and national decline necessarily ensued.[58] On the left, Shapurji Saklatvala (the Communist MP for Battersea North), in discussing naval ratings, stated that he wanted them "to cease being a mere Robot and become a human being."[59] Wilfred Wellock lamented the fact that workers had become appendages of the machine, their intelligence eliminated. In wartime, blind obedience was naturally sought; more disturbing was that in peacetime, industrialists, in seeking product uniformity, allowed machines to dictate to men.[60]

C. E. M. Joad, the popular Birkbeck professor of philosophy and BBC broadcaster, warned that mass education, film, radio, and tabloid newspapers were homogenizing the population and destroying individualism. And now came the machine. "There is real danger that modern industry will substitute the robot for the craftsman." The skilled laborer was disappearing. "Today his type is too often that of the worker ant endlessly performing the same operation in the ant heap factory."[61] In hailing the work of the Tavistock Clinic, the psychotherapist Dr. H. Crichton-Miller concurred that civilization was sick because "creative ability was being killed by the repetitive, the craftsman replaced by the robot."[62] But according to J. B. Priestly, in a cheap and monotonous culture it was not only work that was standardized. "The danger is, of course, that this robot employment will alternate with robot leisure, passive amusement as standardized and impersonal as the tasks of the machine."[63]

What would be the inevitable result of the production of so many unskilled drones? Experts on the history of science fiction have noted that the "'submerged-nation' theme, coupled with the idea of retribution, is essentially a British obsession."[64] Such preoccupations were quite apparent in the discussions of the subjection of laborers to mechanization. The central fear of its opponents was not simply that factory work was reducing workers to the level of robots, but that some day these robot-workers would revolt.[65] What form would the revolt take? In *R.U.R.*, the robots ultimately take up arms against their human masters, but what makes the story so chilling is that they have already won the reproduction wars.[66] Commentators have missed this crucial linkage of the technological and the biological. When talking about robots, contemporaries did not simply discuss machines; they almost inevitably broached the topics of gender, sexuality, and reproduction, and they generally followed the eugenicists in their thinking. On the one hand the advocates of eugenics were all in favor of machinelike efficiency and sought to shape the race and streamline the population according to the dictates of medical science. (Indeed some accused the eugenicists of welcoming a society of robots.) On the other hand they were the very ones who raised the alarm that their increasingly urbanized and industrialized society was being swamped by the biologically unfit. The fear of swarming robot hordes can, indeed, be best read as the middle classes' fear, not of machines, but of the proletarians who tended them in the Fordist factory.

Though warmly reviewing *R.U.R.*, James Agate, London's leading theater critic, made the curious complaint that the author unnecessarily dragged in "the irrelevant ornament of sex."[67] Stating that he would have preferred it if there had been no women at all in the play, Agate revealed that he missed its central point. It was the very thought that the automaton might succeed in reproducing that raised the truly frightening prospect of it freeing itself from dependence on humans. Mary Shelley had Victor Frankenstein realize in time that if he succeeded in producing a mate for his creature, the consequences would be horrific. *R.U.R.* ends on a more hesitantly positive note, but throughout the play the vital importance of reproduction is underscored. The central argument in *R.U.R.* is that the creation of the robot led to a crisis in human reproduction. With machine men doing all the work, real people have stopped reproducing. Dr. Gall explains that ever since surpluses in production occurred, the birth rate has dropped. Despite knowing the disastrous demographic consequences, politicians in search of power and industrialists in search of profits stupidly persist in ordering more and more automatons. In the week before the revolt breaks out, the newspapers report that not a single human birth has been recorded. In explaining why she finally destroyed the formula for the production of robots, Helena protests, "If the manufacture of the Robots had been continued, there would have been no more children. . . . Everybody said that human beings could not be born because so many Robots were being made."[68]

This was exactly the sort of demographic crisis that eugenicists were constantly harping on—the fact that the fit were having small families and the unfit large broods.[69] When the statistics documenting such patterns were published, cries went up that something had to be done. George Bernard Shaw complained, "Being cowards we defeat natural selection under cover of philanthropy; being sluggards, we neglect artificial selection under cover of delicacy and morality."[70] This argument was being made decades before the appearance of *R.U.R.* The first volume of the *Eugenics Review* (1909–10) carried an article by Dean Inge arguing that race improvement depended on deciding who breeds. "Rational selection" was now needed since nature no longer curbed the threatening fertility of the urban proletariat. Degenerates needed to sacrifice themselves, while it was the duty of good stock to breed.[71] According to the Whethams, birth control was limiting fertility, but the issue was one of quality rather than of quantity. If middle-class women shunned maternity, they posed a "direct menace to the future welfare of the race."[72] A feminist like Helena Swanwick agreed that the finest women were less prolific than the most degraded, and the feebleminded the most prolific of all.[73] Sterility, she believed, was due to idleness and luxury on the one hand and poverty and destitution on the other.

In the 1920s, the idea of a crisis in reproduction was very much in the air. Modern, mechanized society was held responsible for producing both shoddy products and unfit families. This idea that the worst sorts of people were reproducing while the best were not, that the excessive fertility of the unfit was in effect sterilizing the fit, was repeated endlessly, especially by those hostile to birth control. Aldous Huxley had a character in his novel *Antic Hay* (1923) express the common sentiment: "It is regrettable that birth control should have begun at the wrong end of the scale. There seems to be a level of poverty below which it doesn't seem worth while practising birth control and a level of education below which birth control is regarded as morally wrong."[74] Science-fiction writers sketched out the likely catastrophic consequences. In "Automata" (1929), Sydney Fowler Wright portrayed a late twentieth-century world in which machines are everywhere, human procreation has almost ended, and the robot replaces the "flesh-child."[75] "If a machine proves to be sufficiently useful," Edward J. O'Brien noted sardonically, "it is not allowed to practise birth control. It reproduces itself as rapidly as possible until a better machine of its own type consigns it and its likes to the scrap heap."[76]

Birth controllers themselves worried about differential fertility. Marie Stopes argued that in the past weaklings died off. The effect of modern state welfare had, in her words "been to work havoc with the quality of the nation." The fit were taxed by the state to support "the unhappily over-breeding, feeble-minded slum dwellers," and the middle class were in effect "sterilized" by the poor. From a eugenic point of view such movements were disastrous. "Soon the only class callously and carelessly allowing themselves to hand on bodily defect will be the morons of the various grades, sometimes called the 'social problem group.'" Stopes's answer was to take birth control information to the poor. "What stands out as new and therefore as essentially characteristic of the time is the rational application of scientific thought to the problems of marriage."[77]

The eugenically minded continually stressed that men and women had distinct biological roles and any blurring of the boundaries could be demographically dangerous. Given the British preoccupation with the crisis in reproduction, it is understandable that they would respond to *R.U.R.*, a heavily gendered play. All the managers are men, and Helena Glory is the only woman on the island. The *Spectator*'s reviewer made the good point that Helena is a curious character. She is in no sense the assertive 1920s flapper, yet in remaining childless after five years of marriage she also fails to fulfill the natural female role.[78] Curiously enough some of the female robots are more traditionally feminine than she. Helena's first act upon arriving on the island is to mistake Sulla, a "Robotess" secretary, for a human. When she asks Domaine why female robots are produced, she is told that customers expect domestic services to be provided by females and the firm simply responds to a demand.

Gender stereotyping was common in such literature. Mrs. M. L. Campbell's "The Automatic Maid-of-All-Work" (1893) tried to wring some humor out of the idea of a clock-faced, electric-powered servant girl stupidly washing dinner, making the bed with the man of the house still in it, and finally setting about herself with an axe.[79] Nevertheless the assumption was that domestic chores would necessarily be done by a "female" robot. The comic writer Stephen Leacock played up the old line about women's loquaciousness in envisaging female robots being equipped with a vocabulary twice the size of that of their male counterparts.[80]

In Leacock's story, robots are used as go-betweens when shy human couples engage in marriage negotiations. The robots, of course, cannot marry. In *R.U.R.*, male and female robots exist, but they have no love or sexual feelings for each other. Ultimately a special female robot is made by Dr. Gall who resembles Helena in beauty but is lacking in intelligence. This second Helena asks what purpose can she have, not being fit for any work. The answer (which any eugenicists would tell her) is that she will inspire love in her mate and eventually succeed in reproducing.[81]

But to pursue Helena's question: Why are there female robots, or as Allison Muri has put it, why do cyborgs have boobs?[82] The answer is that, whereas male automatons can be portrayed as simply working, the Western world always culturally links females to sex and reproduction. Female robots, in blurring the boundary between the biological and the mechanical, are necessarily regarded as either fetishistically appealing or repellently perverse.[83] The sexualized automatons that predate Čapek's work were envisaged as servicing humans. The female robot provided the ultimate male fantasy of the compliant woman. In E. T. A. Hoffman's "The Sandman" (1815), Nathanael calls his human fiancée Klara a "lifeless damned automaton" and ironically falls in love with a mechanical woman, Olympia.[84] The "Vénus anatomique" in Rachilde's novel *La Marquise de Sade* (1886) is a symbol of female passiveness and exploitation.[85] In *Tomorrow's Eve* (1886), Villiers de l'Isle Adam has Edison produce an android—Hadaly—for a friend repulsed by real women. Though the automaton is completely controlled by men, the novel ends with a twist. In revealing that a female assistant actually uses Hadaly to seek revenge against men, the novel highlights male fears of hybrid females.[86] And just as in *R.U.R.* there is the real Helena and the robot Helena, so too in Thea Von Harbou's *Metropolis* (1927), there are two Marias. Maria the evil robot (in Fritz Lang's filmed version very much the sexually alluring vamp) almost succeeds in seducing the workers into a self-destructive rebellion.[87] More reassuring were the pornographic accounts of completely compliant sex dolls in works like *La femme endormie* (1899).[88] Such beings were not mere fictions. The German sexologist Magnus Hirschfeld provided photographs of commercially produced, inflatable frauleins.[89] Of course men had long sought and were repulsed by the robot-like passivity of the prostitute. In *Against the Grain* (1884), Huysmans has des Esseintes describe the whores of the Latin Quarter who "all, like so many automata wound up at the same time with the same key, uttered in the same tone the same invitations, lavished the same smiles, talked the same silly phrases, indulged in the same absurd reflexions."[90] Such musings on the continuum of prostitute, sex doll, and compliant female were common on the continent. It appears that no early twentieth-century British author wrote explicitly about sexed female robots.[91]

Could a British woman be attracted to a mechanical man? Alongside a story entitled "Are There Robots in Real Life?" written in response to *R.U.R.*, the *Daily Mirror* ran a cartoon showing a woman who buys a robot because, unlike real men, he loves to dance and never complains about her tardiness.[92] But male robots—especially British male robots—were usually not presented as sexually active. In E. V. Odle's *The Clockwork Man*, a cyborg from the fifty-ninth century with a clock mechanism in his head, finds himself in an English village. He explains how, in the future, mechanical aids increasingly provide "some means of supplementing the normal functions of the body." He has something like a heart, but the author categorically states, "The Clockwork man had no apparent sex." And so cannot reproduce. In the future, new beings are produced by the makers who have taken away all the women. When the robot sees human lovers, he wistfully muses, "All that old business—before we became *fixed,* you know. But they had to leave it out. It would have made the clock too complicated. Besides, it wasn't necessary, you see. The clock kept going forever. The splitting up process went out of fashion, the splitting up of yourself into little bits that grew up like you—offspring, they used to call them."[93]

"Robert" the robot, the chief character in Joseph Storer Clouston's *Button Brains*

(1933), has a disastrous career as waiter and department store server. Though he has no sexual appetites, young women find him attractive, and the maid Fanny is discovered wooing him. The comic novel advances the elitist, middle-class view that a deprived servant might well be charmed by the programmed attentions of a machine.[94] Robot courtship could be the basis for comedy; robot reproduction could not.

To counter the perceived crisis in reproduction, eugenicists proposed selective breeding by both positive and negative means. Positive eugenic programs aimed at encouraging the reproduction of the fit. Such views were echoed in *R.U.R.* Helena remains childless while on the island, but believes that if she and Domaine escape, they will have a home and "children will come to us at last."[95] For the eugenicists, it was vital that such middle-class women reproduce. Reminders to fit women of their duty to "breed well" were repeated endlessly.[96]

Countless commentators stressed the need for selective breeding. Shaw believed birth control, in allowing people to weed themselves out, was leading to "intelligently controlled, conscious fertility."[97] His hope was that soon "the great central purpose of breeding the race, ay, breeding it to heights now deemed superhuman," would be recognized, and no longer just a personal fancy.[98] Bertrand Russell foresaw a society of Edisons with the strength of prizefighters.[99] A similar portrayal of breeding a desired type occurred in the work of the eugenicist Reginald Ruggles Gates. Though opposed in general to miscegenation, he enthusiastically reported that white/Indian crosses resulted in "greater initiative and enterprise than the pure Indian would ever show."[100]

A eugenic-minded commentator like the social hygienist Sybil Neville-Rolfe insisted that medical exams before marriage were necessary if one was to know "the hereditary qualities of the stocks that are to contribute to the new individual."[101] In *Vitality: A Book on the Health of Women and Children* (1935), Elizabeth Sloan Chesser, a medical doctor, agreed that natural selection was no longer operating. Soon marriages would be regulated, medical examinations required to wed, and the feebleminded segregated. The state had to encourage the families of "worthy parents" as a high birth rate of the fit was essential for future prosperity.[102] Women were the target of most of these injunctions. Norah March presented eugenics warring against "incompetent motherhood" and the "racial poisons" of alcohol, tuberculosis, and venereal disease.[103] H. G. Wells foresaw women having careers of "wholesome motherhood."[104] The best mothers required the best sorts of social supports. Chesser discussed the necessity of state hospitals, pregnancy insurance (to allow home stays after birth), and mothers pensions.[105] A number of feminist supported such eugenic views.[106] Perhaps the most radical were contained in the futuristic novel *Man's World*, in which Charlotte Haldane envisaged a society dominated by women in which mothers, serviced by mechanical slaves, would devote themselves to childbearing while sterilized "entertainers" dealt with men's lust.[107]

Descriptions of state-supported women happily dedicating themselves to childbearing remind one of the sort of cyborgs later made famous by the 1975 film *The Stepford Wives*.[108] Even in the 1920s commentators concluded that it was likely not possible to force middle-class women to be prolific child producers. J. B. S. Haldane famously satirized such ambitions. "The eugenic official, a compound, it would appear, of the policeman, the priest, and the procureur, is to hale us off at suitable intervals to the local temple of Venus Genetrix with a partner chosen, one gathers, by something of the nature of a glorified medical board."[109] Reactionaries like R. Austin Freeman agreed. "The idea of compulsory mating of human beings is merely fantastic; and it may be added that any individuals who would permit themselves to be 'bred' like pedigree bulls would not be worth breeding."[110] Since such positive eugenic programs were doomed, how then did Freemen propose dealing with the "sub-men" who, composing about one-fifth of the population, acted as a drag on society? The answer lay in controlling their reproduction by segregation, marriage restriction, and sterilization.

R.U.R. ends with Radius, the robot leader, seeking the secret of reproduction. "The human beings did not let us breed," he cries. "We are sterile—we cannot beget children."[111] The robots' complaint would have reminded many that the eugenicists argued that sterilization of the unfit was a necessary means by which to purify the race. Radius's lament would thus have warmed the heart of the biological determinist. If only the "sub-men" were also sterile, there would have been no reproductive crisis. In fact, the eugenicist claimed that due to their lesser intelligence, lack of moral restraints, and untrammeled sexuality, the unfit were dangerously prolific. Therefore some were led to the logical conclusion that it was necessary to sterilize them. Many leading lights of the Eugenics Society entered the discussion. Its general secretary C. P. Blacker and its honorary president Leonard Darwin called for the voluntary sterilization of the feebleminded. Although not all eugenicists supported passage of sterilization legislation, the movement and the policy were always inextricably entwined in the public mind.[112]

In 1892, S. A. K. Strahan protested that the unfit, no longer weeded out as in nature, were being allowed to produce "wretched offspring." More care was taken in breeding cattle. Legislation was needed to control their reproduction.[113] Not surprisingly, a reactionary like Anthony Ludovici defends the need for selective culling.[114] In a radio broadcast, Sir James Jeans, the Cambridge astronomer, agreed that science had to prevail over humanitarianism. "If we are to make the earth a paradise again, it seems to me that our first duty is, at all costs, to prevent the moral, mental and physical wreckage of today from reproducing itself and starting a new sequence of unhappy lives trailing down through endless generations. To encourage this stream of misery becomes a vice; to check it a virtue."[115] Socialism was, according to Jeans, a disease. Breeders knew that qualities were inherited, and no environment could improve a weakling. Fortunately there was now a reaction against the old sentimentalism by "the rising generation," but the real attack against the incubus of the defective was still to come.[116]

Women writers entered the fray. In a chapter titled "Motherhood and Eugenics," Elizabeth Sloan Chesser argued that the feebleminded had to be either segregated or sterilized. To do nothing would be tantamount to demographic "suicide."[117] Writing in the 1930s Sybil Neville-Rolfe argued that marriage was not a right but a "privilege." Sterilization in Britain for eugenic reasons, she insisted, was not to be compared to its employment in Nazi Germany where it was turned to political purposes.[118]

Some progressives supported such a culling. The prominent Labour Party activist Frida Laski was a member of the Workers Committee to Legalize Voluntary Sterilization. The iconoclastic scientist Lancelot Hogben was not opposed to the sterilization of criminals.[119] John Desmond Bernal, Cambridge researcher and sometime member of the Communist Party, presented sterilization as a form of social self-defense.[120] The philosopher Bertrand Russell predicted that eventually the state would determine the sorts of minds and bodies it wanted. Under the tyranny of science, few would be needed as breeders; most would be sterilized.[121] According to the child psychotherapist Susan Isaacs, many young people were handicapped "by the results of vices practised by parents in carelessness or ignorance." Recent reports confirmed her suspicion that "there has been much breeding from bad stock and that this is one reason for the ever-growing number of our defectives."[122] Claud Mullins, a magistrate who saw many unhappy marriages and overlarge families, argued in 1933, "Surely the time has come for at least an earnest consideration of the whole question of the right of the unfit to marry and propagate." Sterilization had to be carefully considered, and the poor mother with a brood of children "ought to be brought to realize the moral crime of having more babies."[123]

Aldous Huxley's "A Note on Eugenics" held that a majority of humans were born "dull-witted" and so would inevitably be ruled by the intelligent and strong. Unfortunately the inferior sort was increasing in number while better types were not. His modest proposal was for the sterilization for the feebleminded and bonuses for the bright. Though he claimed not to be arguing for eugenic reform, he painted a nightmare scenario of the coming of an era of "sub-men and super-men." He cited Leonard Darwin as noting wage earning and fitness were correlated so it was easy to see whose fertility to support and whose to curb. Yet Huxley believed an underclass of automatons would always be needed. "States function as smoothly as they do, because the greater part of the population is not very intelligent, dreads responsibility, and desires nothing better than to be told what to do. . . . If the eugenists are in too much of an enthusiastic hurry to improve the race, they will only succeed in destroying it."[124] Like so many of his contemporaries, Aldous Huxley was of two minds about eugenics, condemning its social conservatism while defending the sterilization of "nitwits" who were "at large."[125]

R.U.R. could not only be read as a eugenic text, reviewing positive and negative eugenic strategies, it also echoed a current interest in the most radical form of medical intervention associated with the advent of a secular, scientific society—euthanasia. In the play, the robots usually live for about twenty years, but if they prove defective, cramp up, or go off their heads, they are sent to the stamping mill to be destroyed.[126] Čapek thereby reminds us that the discussion of eliminating useless humans was also being broached in the early twentieth century. Following Dr. Benjamin Ward Richardson's patented "Lethal Chamber for the Painless Extinction of Lower Animal Life," in 1884 the Battersea Dogs Home began to employ carbonic acid gas to painlessly terminate the lives of animals.[127] As a result, the term "lethal chamber" was soon in popular use and, suggestions were made that it be used on humans.[128] Huxley included joking references to the "lethal chamber" in *Crome Yellow.* When Scogans, the scientist, is asked by Denis, the poet, if he will have any place in the future rational society, he is told, "No, I can see no place for you; only the lethal chamber."[129] D. H. Lawrence fantasized about the uses to which he would put such a killing device.[130] Conservatives suggested that only in the Soviet Union could people seriously entertain such ideas. A correspondent of *The Times* asserted that "one fails to see how they [Bolsheviks] can regard themselves or other men as anything better than 'robots,' machines without a soul, or a God, or a hereafter, without gratitude or love or pity, pieces of mechanism that should, if incurably inefficient, be sent to the stamping-mill or the lethal chamber."[131]

Nevertheless in Britain the eugenically minded were interested in any method that could be used to purge the race of the defective. In *A Modern Utopia,* H. G. Wells argued that in time the citizenry would come to accept the death of the diseased or helpless as "an act of love." In the future society, he predicted, "There would be no killing, no lethal chambers. No doubt Utopia will kill all deformed and monstrous and evilly diseased births, but for the rest, the State will hold itself accountable for their being."[132] Carveth Read, Grote Professor of Philosophy at the University of London, declared himself "consoled" that the lower classes had a high death rate. He concluded, however, that nature had to be assisted in the selection process. "Whether anything can be done to weed the population by permanently segregating criminals and quasi-lunatics and imbeciles, by preventing marriage amongst certain classes of invalids and hereditary suspects, by euthanasia of certain idiots and incurable sufferers, time will show."[133] George Bernard Shaw shocked the London Eugenics Education Society by a 1910 lecture in which he crudely suggested that members of the society were exploiting the taxpayers' grubby desire to spare themselves the cost of caring for the unfortunate. "A part of eugenic politics would finally land us in an extensive use of the lethal chambers. A great many people would have to be put out of existence simply because it wasted other people's time to look after them."[134] Shaw was being his usual provocative self and attempting to satirize such views, but the public took it as further evidence of the eugenicists' desire to eliminate the unfit.[135] Such associations were strengthened by books like *The Survival of the Unfittest* (1927) in which Charles Wickstead Armstrong included the lethal chamber along with sterilization and segregation as ways to limit the fertility of the "scum."[136] The same author envisaged a future utopia called "Eugenia" where the unfit would happily avail themselves of the services offered by "Euthanasian Gardens."[137] The Eugenics Society shunned such enthusiasts, but when the British Voluntary Euthanasia Society was founded in 1935 it did include such well-known eugenicists as C. Killick Millard, George Bernard Shaw, H. G. Wells, and Havelock Ellis.[138] The journalist Philip Mairet maintained that class concerns still marked such discussions. The middle class continued to be frightened by the fertility of the lower classes. "*They* don't want posterity entrusted to the labouring classes—they begin to splutter at the very idea, begin talk about sterilization of the unfit (even lethal chambers have been mentioned)—at the very least they would teach the lower classes how to limit their families."[139]

In *R.U.R.*, no matter what Domaine tells Helena of the robots having no attachment to life, she continues to be horrified at the thought of their being consigned to the stamping mill. Does knowing that the eugenically minded were in the first decades of the twentieth century broaching the issue of euthanasia help in appreciating the response to the play? It does on a number of levels. It first suggests that in sketching out his drama, Čapek drew on a topical issue. Second, it alerts us to the likelihood that the play's audience was divided on the issue. Although the vast majority probably identified, as they were expected to do, with Helena's point of view, there were no doubt some who would have necessarily shared Domaine's argument—echoed by the eugenicists—that the rational had the right to eliminate beings who lacked both reason and social worth.

Up to this point we have mainly traced the ways in which eugenics, by warning the nation of dangers of the fertility differential and in proposing such remedies as sterilization of the unfit and the elimination of the defective, positioned itself as a bulwark against hordes of robot-like workers. But the opponents of eugenics turned the robot analogy to exactly the opposite purpose, arguing that it was the eugenicists, with their mania for calculating the quantitative and qualitative value of human life, who were clearly most enamored by the automaton.

Early on, those speaking for the working class recognized the ambition of some scientists to produce man-machines. An editorial entitled "Life as per Recipe," in the *Daily Herald* of September 6, 1912, warned of a world where workers would be produced,

> in a nice, clean laboratory, under Government supervision and control. (This sort of thing is always under supervision and control.) Nice, well-intentioned scientists will successfully perform, in fact, what Frankenstein achieved only in fiction. Can't we hear them talking? "Better give this one a little more bicep: looks as if coal-miners are running short." Or "You're allowing that man too much brain space; he'll be joining a trade union if you don't take care." Or "Are you sure he won't be able to have children? It'll be confoundedly inconvenient if he does.". . .
>
> These imaginary scraps of conversation are not so imaginary; for what purpose is the new discovery most likely to be utilised if it can be carried out to its logical conclusion? To the making of men, to the making of national backbone? No. The new discovery will be exploited, as every new discovery of modern science has been exploited, immediately contrary to the best interests of the nation. Elderly duchesses will be kept alive, their pug dogs will have their dying hours delayed, and the greatness of science poured out upon the least worthy.
>
> As to the workers . . . it is easy to see what may happen. The life creators will join with the eugenists. A section, and probably a large section of the working class will be forbidden to have children. . . . The necessary population, calculated in Whitehall, will be supplied with all their virtues and vices—or shall we say virtues, and leave it at that?—ready made. The slave foredoomed will always be a slave; the scientists will have seen to that.[140]

Some feminists countered the eugenicist vision of women dedicating themselves in robot-like fashion to childbearing. Helena M. Swanwick noted that eugenists might say who would make the best mothers, but the woman of future would have her own ambitions. Reactionaries talked of the girl needing to store up energy for motherhood "as if a human creature were nothing but a chemical factory and warehouse rolled into one." Such training of women "as breeders only," following the "theory of the cow-woman, who shall do nothing but bear and suckle babies," led to subjection. For Swanwick, such ideas grew out of the industrial world's mania for an unhealthy division of labor.[141]

Scientists objected that the eugenicists' desire to purge the nation of the abnormal and impose new standards of conformity could also have the unexpected consequence of racial stultification. Sir Humphrey Rolleston, Regius Professor of Physic at Cambridge University and President of the Royal College of Physicians, warned his colleagues: "Without being reactionary we may wisely hesitate before advocating strict eugenic measures of breeding, which, if carried to their logical conclusions, might seriously impair the future progress of the race; for if the inborn tendency to variation, which is responsible both for mental weakness and for intellectual ability, were thus removed, a dead level of standardized men, like 'Robots,' might conceivably result."[142] A similar line was taken by Aldous Huxley, when he chose to be critical of eugenics. "According to Major Leonard Darwin, the fittest to survive are those who can earn the most money. The deserving rich must be encouraged to propagate their kind; the poor, whatever the cause of their poverty, whether it be illness, eccentricity, too much or too little intelligence, must be discouraged and if necessary sterilized. If Major Darwin gets his way, the world in a few generations will be peopled exclusively by Podsnaps and Babbits."[143]

Eugenicists were held responsible for seeking the homogenization of society. There would be, many agreed, no place for erratic geniuses if the eugenicists succeeded in establishing a "race of standardized types."[144] When Anthony Ludovici argued that the unfit should be prevented from marrying, he was attacked in the press. According to Stanley L. Baker, only if human nature changed could such zealots "by a careful selection of species, produce a nation of robots."[145] In an article titled "Shall We Get a Race of Human Machines?" John Malcolm reprimanded medical experts for seeking to predict children's abilities. The more science intervened in child rearing, he believed, the more likely one would have a generation of robot children who never thought for themselves.[146]

When the 1932 National Radio Exhibition ended, the *Daily Mirror* reminded its readers that the most popular item on display had been "Alpha" the robot. A journalist captured the public's ambivalent view of the automaton by honoring it in a poem.

When the future comes to birth, Ruthless Robot,
Will your brothers rule the earth?
Will the Robot
Hold mechanical control
Of the world from pole to pole?
What of man's immortal soul?
Ask the Robot.[147]

Many of the reasons for the interest in the robot are fairly obvious. For many in Britain, uncertain of the benefits of mechanization, the mechanical man posed the possible threat of displacing the human worker. What is less obvious is that the robot story enjoyed added resonance because it could also be read as a eugenicist parable. It is no exaggeration to say that when the British were talking about automatons they were consciously or not talking about laborers and how they might best be controlled. Robot stories and eugenic accounts told much the same tale. They both highlighted the promises and threats posed by modernity. The harnessing of science and industry, they argued, could lead to unparalleled progress, but if not directed by an enlightened elite with the interests of the race at heart, the unleashed forces of rationalization would result in disastrous consequences. And if production was carefully planned, there was all the more reason to see that human reproduction be similarly designed. Contemporaries had good reasons for regarding eugenics as either the best hope of countering the population problems modernity had created or the worst example of the age's penchant for technological fixes.

PART TWO

Beyond the Predictive

Sex in Real Time

CHAPTER 4

"A sort of animal or mechanic immortality"

Glands and Eugenics

Aldous Huxley's *Brave New World*—best known for its account of raising babies in bottles—is replete with descriptions of a future society's use of hormones. It is portrayed penning up thousands of cattle to provide both gland extracts and milk to an Internal and External Secretion Trust. Its scientists stimulate embryos with placentin and thryoxin, corpus luteum extract, doses of pituitary, hog' stomach extract, and fetal foal's liver. They make some female embryos into freemartins (infertile females) by treating them with a male sex hormone. As adults, women take pregnancy substitutes—syrup of corpus luteum, ovarin, mammary gland extract, and placentin—while men chew sex hormone gum and exchange panglandular *petits beurres*. All citizens have monthly "Violent Passion Surrogate" treatments as they are told that stimulated adrenals are needed for good health. The regime so arranges it that the individual can thus experience the tonic effects of fear and rage while society is spared any accompanying disturbances.[1]

In portraying so imaginatively the ways in which endocrinology might be exploited by the state, Huxley was tapping into contemporaries' fascination with the hormones—those mysterious, recently discovered, and apparently powerful substances. Produced by the ductless glands, they promised on the one hand to reinvigorate the elderly and on the other to cure the sick and calm the troublesome. Optimists hoped that by their employment society would have the means of dealing with many of the physiological and psychological disturbances exacerbated by the rise of modernity. The less sanguine feared that instead of restoring a "natural order," scientists and doctors who tampered with hormones might render ever more fragile the boundaries that separated the sexes, and indeed the species. Both sides recognized that endocrinology promised to provide eugenics and sexology with the means of improving human efficiency, streamlining the population, and avoiding waste of energy. These were not unintended consequences. Many of those carrying out hormonal research explicitly proclaimed the important part the glands could play in the campaign to achieve reproduction by design.

Hormonal therapies—along with organ transplants, artificial insemination, and prosthetics—provided the most startling evidence possible of modern medical scientists' abilities to repair or possibly even remake men and women. But in the interwar years, few laypersons or scientists actually employed the terms "endocrinology" or "hormones." What they did discuss was the subject of "rejuvenation." We will accordingly begin by reviewing the history of this therapeutic process, before turning to those who debated the efficacy and morality of exploiting the ductless glands for sexual and social purposes.

In a 1924 review, F. A. E. Crew, animal geneticist and member of the Eugenics Society, observed that researchers' reference to "rejuvenation" had for some years now provided comedians with jokes, journalists with headlines, and dramatists with scenarios. It was time for a scientist to clarify the issue. The public wanted to know if aging could be retarded and death postponed. What role did the glands play? Could repairs be made and rejuvenation won as claimed by those who used transplants, injections, and ligation of the vas deferens? Crew conceded that the evidence based on animal experiments suggested it was possible. Injections he considered useless, testicular grafts (from either ape or human donors) were difficult to obtain, but the so-called Steinach operation, consisting simply of the ligation of the vas deferens, seemed to energize male patients, while claims were made that irradiation of the ovaries by x-rays (though causing sterility) brought similar benefits to women.[2]

The age of the hormones had dawned. But what exactly were they? During the famous 1922 Russell divorce trial, a fortune teller bizarrely claimed to have diagnosed Christobel Russell's pregnancy by "the vibration of the hormones."[3] Gilda, in Noël Coward's comic play *Design for Living* (1932), when asked to explain her erratic behavior, replies: "Glands, I expect. Everything's glandular. I read a book about it the other day. Ernest, if you only realized what was going on inside you, you'd be bitterly offended. . . . All the hormones in my blood are working overtime. They're rushing madly in and out of my organs like messenger boys."[4] The general public assumed that almost every scientific reference to "glandular secretions" was likely related in some way to sex. When Julian Huxley's research on thyroid gland extracts to accelerate the growth of salamander-like creatures was publicized, he was immediately badgered by people looking for a sex stimulant.[5] Such assumptions were understandable given that in the 1920s and 1930s the first time most laypeople read about hormones was in sensational reports of sexual rejuvenation.

For obvious reasons, the earliest research in endocrinology had focused on the sex organs. As far back as the Greeks, it had been known that the testes were responsible both for a man's fertility and for his secondary sexual characteristics. In the 1800s, the question began to be asked exactly what the gonads did aside from simply produce gametes. That the testicles had both a reproductive and an endocrine function was dramatically demonstrated in 1849 when A. A. Berthold, in transplanting testicular tissue, restored male sexual behavior in castrated roosters.[6]

Medical scientists' working assumption was that men and women produced antagonistic "male" and "female" glandular fluids. To be more "manly" accordingly required obtaining in some fashion the magical masculine substance. Such a belief underpinned the work of Charles Edouard Brown-Séquard, the British-born pioneering neurologist, who created a sensation in June 1889 by asserting that he had carried out testicular experiments on human subjects including himself. In his report to the Société de Biologie in Paris, the seventy-three-year-old scientist asserted that castration, masturbation, and all other forms of seminal losses resulted in physical and mental debility. Semen obviously had to contain some special strengthening substance—a "dynamogenic power." To test this theory, Brown-Séquard subcutaneously injected himself with a solution of water mixed with the blood of the testicular veins, semen, and juice from the crushed testicles of dogs and guinea pigs. Claiming that as a consequence his bodily strength had been miraculously restored, he strenuously denied that the results were due to autosuggestion.[7] Despite the scientific community's skepticism, a number of enterprising medical men in Europe and North America followed up on Brown-Séquard's assertions in producing for "weak men" a variety of glandular extracts such as "spermine." Though many cures were claimed, such foreign testicular material could have been of no value given that the male hormone was produced—not stored—in the testes and was not water soluble. It made no difference. Brown-Séquard had launched a craze for organotherapy.[8]

The hopes that such procedures might work were kept alive by real advances made in some forms of glandular therapy. It was discovered in the 1890s that sheep thyroid injections cured hypothyroidism.[9] And though Brown-Séquard's claims of having discovered a rejuvenating elixir proved wrong, doctors came to realize that the ductless glands did play a key role in sexual development and a variety of other physiological processes.[10] In 1905, E. H. Starling coined the term "hormone" to refer to the chemical messengers produced by the various ductless glands. Such was the assumed importance of the hormones that Sir Arthur Keith described them as providing the body with a "system of government" that gave orders to the various subservient organs.[11]

The best-known sex gland experiments in the early twentieth century were carried out by Serge Voronoff (1866–1951), an eminent Russian-born medical scientist working at the Collège de France. Beginning his career experimenting with skin and bone grafts, he was led on to embrace the notion that the aging process could be reversed by transplanting the testes of young animals into the old. If a man's sex glands were decayed, so too were the "moral and physical energies" of manhood. In 1919, he scandalized many by transplanting the testes of young chimpanzees into elderly men, asserting that "marked psychical and sexual excitation" typically resulted, followed by a resurgence of memory, energy, and "genital functions."[12]

The British reading public avidly followed Voronoff's work on what were commonly called the "monkey glands." In 1922, he presented to a Parisian audience an English patient who testified to the value of the operation.[13] The *Lancet* carried articles examining the legal complications of testicular transplants and went so far as to suggest that large hospitals might establish "a sort of exchange and mart" in organs.[14] In 1924, Voronoff lectured in Edinburgh, bringing with him a film documenting both rejuvenated rams and humans. Most members of the British medical profession were not convinced. In an otherwise sympathetic lecture on "Monkey Glands and Others," Professor Winifred Cullis of the obstetric unit of the Royal Free Hospital warned that public expectations of extending life expectancy should not be too high. "If what we desired was to increase the span of life," she cautioned, "we would have to turn to eugenics."[15] Nevertheless medical journals continued to publish articles supporting Voronoff's work, and the editors of the 1926 edition of the *Encyclopaedia Britannica* invited him to produce an essay on rejuvenation.[16] George Bernard Shaw reported that he had been approached by Voronoff but had no desire to employ monkey glands to prolong his own life.[17] Nevertheless thousands of such expensive operations were carried out on wealthy men, including Maurice Maeterlinck, the Belgian poet and Nobel laureate.[18] Around 1928 Voronoff's popularity in Britain peaked, with his being feted in Cambridge and London.[19]

Unlike Voronoff, the Viennese physiologist Eugen Steinach (1861–1944), was not so much interested in exploiting the therapeutic potential of the ductless glands as in understanding their role in the process of sexual development. He discovered that by transplanting gonads the sexual characteristics and behavior of the guinea pig could be changed, though not its sex. He deduced that in males the seminal vesicles contained two types of cells, the germinal cells that produced sperm and the Leydig or interstitial cells that produced an active substance that, in entering the blood stream, fueled sexual comportment.[20] And when vasectomies were carried out on rats, Steinach reported, the result was an atrophy of the testicle's seminal cells, a proliferation of the Leydig cells, and subsequently an increase in sexual activity.

In 1918, Steinach asked the surgeon Robert Lichtenstern to test the procedure on a human subject. The beauty of Steinach's operation consisting of the unilateral ligation of the vas deferens—that is, cutting only one of the two vas deferentia—was that the patient, though not sterilized, supposedly received a hormonal boost from himself. To achieve the latter effect, American doctors had employed the more unseemly methods of harvesting the human testicles of either accident victims or executed criminals while Voronoff used the tissues of primates. Steinach declared his operation was successful, though primarily in the sense of making men—not younger—but more productive.[21] Steinach's chief concern was not with proper sexual function, but with a more general sort of well-being and vitality. In his words, "sexuality in its various expressions is the measure for, not the purpose of, reactivation."[22] Believing his procedures could alleviate a host of ills ranging from impotence and premature ejaculation to tuberculosis and cancer, his most devoted followers likened the importance of rejuvenation to Darwin's theory of evolution.[23]

Steinach's operation received almost as much attention in Britain as did Voronoff's more sensational transplants.[24] Medical journals like the *Lancet* kept professionals up to date on his experiments.[25] References to his operation were made in high and low culture. In Aldous Huxley's *Antic Hay* (1923), a scientist's lab is described as containing a cock engrafted with an ovary, beetles with transplanted heads, and a "fifteen-year old monkey rejuvenated by the Steinach process."[26] In 1929, the *Daily Mirror* carried a story of a decrepit Camden dog rejuvenated by the Steinach operation.[27] Between the 1930s and the 1950s chlorophyll tablets sold as "Phyllosan" that purportedly increased the vital forces of the middle aged were advertised by such slogans as "Fortifies the over-forties" and "Rejuvenation without Operation."[28] Thousands of Steinach operations were carried out in Europe and North America. Mr. Alfred Wilson, a satisfied patient who paid seven hundred pounds for the service, booked the Albert Hall in London for

May 12, 1921 to speak on the subject "How I Was Made Twenty Years Younger." Unfortunately he dropped dead just before his presentation.[29]

Despite such hiccups, Steinach was hailed by a number of sex reformers. In Britain, his most vocal defender was the Australian-born gynecologist Norman Haire. Practicing in London from 1919 onward, Haire was a well-known birth control activist and sexologist.[30] He first heard of rejuvenation in 1921 in a paper read by Eden Paul, a socialist physician, writer, and translator. In 1923, the two men produced a pamphlet on the Steinach operation for the British Society for the Study of Sex Psychology.[31] Haire proceeded in 1924 to devote an entire book to explaining rejuvenation, in which he reported having carried out twenty-five Steinach operations himself.[32]

At the 1929 International Congress of the World League for Sexual Reform meeting in London, the German doctor Peter Schmidt reported successful results in six hundred rejuvenation operations.[33] The left-wing activists Eden and Cedar Paul translated Schmidt's best-known work, *The Conquest of Old Age,* which appeared in 1931. Aimed at popularizing the Steinach operation, the study included a long description of Norman Haire's treatment of a retired fifty-seven-year-old doctor whose decline in sexual potency occasioned "some (physical) disharmony in the relations with the wife, who is a great many years younger." In 1923, he was operated on, and by May his erections were more vigorous, by June his friends said he looked "wonderfully well," and by November he reported the end "of the before mentioned physical disharmony between myself and my wife." In 1927, another decline led him to accept injections of a testicular extract that restored his sexual power. In 1929, Schmidt reenergized him a third time by engrafting him with a forty-year-old man's testicle.[34]

Dr. Kenneth Walker, another pioneering British sex researcher, was more skeptical than Haire about such operations, pointing out that as the testis had to interact with the other glands it was difficult to determine how its hormonal influences might be isolated.[35] Having by 1924 carried out four such rejuvenation operations himself, he concluded that the results could be capricious. He noted that the questions of how one defined senility and how one might eliminate the factor of suggestion had yet to be tackled. But the main barrier to progress, declared Walker, was that the rejuvenation procedure was being financially exploited by some and so was naturally held in derision by most cautious surgeons.[36] The Steinach operation, he reported in 1930, had "come to be regarded as a 'stunt' by the greater part of the profession, and a method of treatment that if used properly might have taken its place in therapeutics has fallen into disrepute."[37]

Sigmund Freud and W. B. Yeats were the most famous patients to undergo the

Steinach procedure. Freud's operation on November 17, 1923, of both vas deferentia was carried out in hopes it would alleviate his cancer of the jaw. He thought the operation had no effect. Yeats believed his sexual and creative powers were restored. His surgery, done by Norman Haire in 1934, won the poet the sobriquet of the "Gland Old Man."[38] The daring young journalist Ethel Mannin later reported that Haire introduced her to Yeats to test the operation's success.[39]

If later endocrinologists tried to disassociate themselves from the practitioners of organotherapy, it has to be stressed that in the 1920s Steinach was highly regarded in the scientific community and indeed nominated several times for the Nobel Prize. Moreover in the decades he and Voronoff were garnering newspaper headlines, initially less well-known endocrinologists were working their way toward the identification and isolation of a number of hormones. Banting and Best's discovery of insulin in 1922 reenergized interest in hormonal research, and finally gave respectability to the science of endocrinology.[40] Experts had high hopes that other diseases aside from diabetes could be conquered by hormonal treatments, yet admitted that though some endocrine therapies seemed to work, they were still not fully understood.

British popular opinion was always guarded in its response to the claims made by the rejuvenators. One observer later noted that "much of the hostility towards rejuvenation has been engendered through its association with sex. . . . The idea has been mooted abroad that those seeking rejuvenation are the immoral, the debauched, the libidinistic, the lecherous members of society; that the treatment prescribed appeals to men who have led dissolute lives, and are in search of a means whereby they can continue their careers of lust and excess."[41] Voronoff's operations moreover easily lent themselves to spoofing.[42] A cocktail called the "Monkey Gland" appeared in the 1920s, consisting of gin, absinthe, orange juice, and grenadine. Bertram Gayton's comic novel *The Gland Stealers* (1922) presented a ninety-five-year-old who is inspired by Voronoff's example to organize an African raid to obtain gorilla glands.[43] The 1926 farce "Do Be Careful" had a plot that a critic noted "suggests 'monkey gland.' The heroine is supposed to have been rejuvenated successfully," but the plot is not.[44] And in the 1934 British film *Evergreen* a character jokes that a young singer looks so much like her famous mother that the audience might believe the latter took "monkey glands."[45]

Voronoff's use of animal grafts naturally garnered the greatest attention. In 1927, Norman Haire stated that the Russian had even transplanted a human uterus into a monkey, fertilized it, "and, at the moment of writing, the birth of a human child from the simian mother is eagerly (if somewhat sceptically) awaited."[46] British culture was both fascinated and appalled by such accounts of the erosion of the human/animal boundary. In 1928, a meeting protesting Voronoff's operations was held at Caxton Hall, chaired by Miss Lind-af-Hageby, honorary director of the Animal Defence and Anti-Vivisection Society. She asserted that the fact that Voronoff's cult had the backing of three hundred surgeon-disciples showed how low medicine has fallen. The Duchess of Hamilton moved a resolution decrying such transplants as "an offence against morality, hygiene and decency," and a violation of the Cruelty to Animals Act of 1876.[47] Yet Voronoff's interests were not as eccentric as they might first appear. His colleague Alexis Carrel, who also worked on animal transplants, was awarded the 1912 Nobel Prize for Medicine for his research on suturing blood vessels, and in 1923 Voronoff shared the stage at an international medical congress in London with Charles Banting who had just received the Nobel Prize for his codiscovery of insulin.[48] The idea of using animal organs as interchangeable machine parts was at the time being voiced in the United States.[49] In the Soviet Union, actual attempts were made at cross-breeding humans and primates.[50]

Some were alarmed at the notion of tampering with the body and believed that transplants produced sex monsters.[51] Writers of fiction were particularly alive to such concerns. In Arthur Conan Doyle's "The Creeping Man" (1923), a professor of sixty-one (roughly Doyle's age) takes the serum of a langur to rejuvenate himself before marrying a young woman, and becomes apelike. Sherlock Holmes spells out the moral. "'When one tries to rise above Nature one is liable to fall below it. The highest type of man may revert to the animal if he leaves the straight road of destiny. . . . There is a danger there—a very real danger to humanity. Consider, Watson, that the material, the sensual, the worldly would all prolong their worthless lives. The spiritual would not avoid the call to something higher. It would be the survival of the least fit. What sort of a cesspool may not our poor world become?'"[52] A similar warning was made in Lord Dunsany's play *Lord Adrian* (1933). An old duke, to spite his heir, marries a second time and sires a child. But because the son (Lord Adrian) is the result of an ape gland graft, it comes as no surprise (to the audience at least) that as a youth he avoids games and has an uncanny love of the forest. He particularly loathes hunting as a manifestation of slavery and slaughter. The domination of nature by man he declares a disaster, every invention "is a step away from the woods, is a step away from Nature, is a step that is wrong."[53] This drama predictably ends when the young man, leaping Tarzan-like from tree to tree in an attempt to rally his beloved forest creatures, is shot dead by the family gamekeeper. The devastated duke belatedly realizes that the tragedy was all due to his tampering with animal glands.[54]

Noël Coward saw the comic possibilities offered by the discussion of gland transplants. In *Private Lives* (1930), he has a couple present their views:

ELYOT: Would you be young always? If you could choose?

AMANDA: No, I don't think so, not if it meant having awful bull's glands popped into me.

ELYOT: Cows for you dear. Bulls for me.

AMANDA: We certainly live in a marvellous age.

AMANDA: It must be so nasty for the poor animals, being experimented on.

ELYOT: Not when the experiments are successful. Why in Vienna I believe you can see whole lines of decrepit old rats carrying on like Tiller Girls.[55]

More important, Voronoff and Steinach were never without their academic critics. The latter pointed out that old age hardens the arteries, which would make rejuvenation dangerous, if it were indeed possible. In addition, as research progressed it became clear that the influences of the glands were all interrelated, so one could not, as Voronoff and Steinach maintained, simply isolate the effect of the interstitial.[56] Fellow scientists further pointed out that Voronoff's claims were particularly difficult to confirm, given that he only reported his successes, never his failures.[57] Professor Starling, who had first coined the term "hormone" in 1905, noted in a 1923 lecture that great progress had been made in ductless gland research (for example, the use of thyroid extracts for cretinism and Banting's insulin to treat diabetes), but attempts at using gonadal extracts had so far failed. Rejuvenation still might work, he admitted, but would not prolong life. "The only way of increasing the duration of life was by eugenics—i.e., by breeding from long-lived individuals."[58] By the 1930s, most experts agreed that transplantation and Steinach's operation either failed or were not proven.[59] Though Voronoff and his patients were convinced that his procedure worked, the scientific consensus was that the body would reject transplanted glands and only quacks could continue to assert the value of such procedures.

For his part, Steinach dropped the ligation of the vas deferens and switched his attention to the injection of glandular solutions. Isolating the male sex hormone posed enormous challenges since the body contains only minute quantities. In 1931, German scientists led by Adolf Butenandt finally succeed in distilling a scant fifty milliliters of the male hormone from the twenty-five thousand liters of urine contributed by the Berlin police barracks. Reviewing *Brave New World* shortly thereafter, Charlotte Haldane dryly observed, "If one wrote to-day a plain description of the work of Dr. Butenandt on the male sex hormone, it would probably seem funnier than Mr. Huxley's detailed opening chapter on the 'Central London Hatchery and Conditioning Centre,' where humanity is 'raised' on model factory lines. But never mind."[60] Supported by the Organon Company of Amsterdam, Ernst Laqueur in 1935 worked out the molecular structure of the hormone that was named testosterone. The same year Butenandt (working for Schering-Kahlbaum) and Leopold Ruzicka (working for the Ciba Corporation) simultaneously reported they had developed the means of turning cholesterol into synthetic testosterone.[61] For their efforts, they were to share the 1939 Nobel Prize for Chemistry. Injections of testosterone did not result in rejuvenation, however, so it remained uncertain as to what exact purposes it would be turned.[62]

Such problems did not phase the commercially minded. From the very start the public's interest in rejuvenation offered the unscrupulous an opportunity. The quacks who exploited the Brown-Séquard affair to sell "Séquarine" competed with a vast range of products guaranteed to fill "weak and nervous men" with "rampant vigour." Phosperine, Damaroids, Osogen, Neurovil, Law Palmetto, Dr. Lecoy's Invigoroids, Gordon's Vital Sexualine Restorative, and the Marston Treatment were only a few of the "Lost Manhood Restorers and Nervous Debility Cures" that the British Medical Association sought to ban.[63]

Physicians who offered hormonal therapies quite naturally feared being seen as quacks. At least one doctor had his name erased from the General Medical Council's register for advertising his offering of the Steinach operation.[64] Another was dropped for unprofessional dealing that included his selling "Hormacton" and "Vitalexin."[65] Nevertheless the medical profession felt pressured by the public and business interests to produce therapies and cures. Well before the identification of testosterone, commercial firms were ahead of the scientific community in claiming that sexual dysfunctions could be remedied by hormone treatments. In the 1920s, the G. W. Carnick Company advertised "Hormotone" for menstrual and climacteric disorders, "Viriligen" for "lowered virility and sexual neurasthenia of functional origin," and an "Orchic-Prostate Compound" for sexual neurasthenia and enlarged prostates. The Harrower Laboratory produced "Gonad Co," a formula mix of adrenal, thyroid, pituitary, plus "spermin" made from Leydig cells for "asexualism" and hypogonadism. The Middlesex Laboratory of Glandular Research claimed to have the records of sixty-four thousand patients, including men who purchased normal and high-potency pills for weak erections and neurasthenia, and women who suffered from frigidity, "sex quiescence," and "inability to reach orgasm." The laboratory's pamphlet for pharmacists advertising "Middlesex Sex Hormones" listed "Emelgemal" tablets for "loss of virility in middle and later life," "Prejacrin" for premature ejaculations, and "Sterility Gland Tablets" for barrenness. One of its window display cards promised "The Fire of Life Replenished," and another asked "Do You Wish to Renew Your Youth?"[66] According to British Glandular Products, "Testrones" could cure men's "impotence, loss of virile tone, weakness, mental and physical apathy, premature aging, sexual indifference, sexual exhaustion, sexual inefficiency, sexual neurasthenia, diminished potency, nocturnal enuresis, sexual functional imbalance."[67] Women could likewise use "Overones" as a natural method of rejuvenation. And finally a 1929 thirty-six-page pamphlet entitled *Rejuvenation of Man and Woman,* which reproduced photographs from Voronoff's books, vaunted the efficacy of "Viriline" for men and "Fertiline" for women.[68]

If the public was fascinated by what it was told of hormones—magical substances, which in the tiniest of quantities could have life-altering effects—social commentators immediately recognized that endocrinology offered the means by which not simply individuals but society as a whole could be refashioned. Indeed Gerald Heard, the science writer and philosopher, claimed that the history of civilization could be explained hormonally. In dogs, the "glands of conflict" were more powerful than the "glands of reflection," that is, the adrenals were bigger than the thyroid; the reverse was true in man. Adrenals prepared one to fight—the blood clots more quickly, the emotions are raised—but man was more influenced by the thyroid that elicited "endurance, patience and receptivity." Due to such flexibility, humans were able to evolve.[69] The well-known public commentator C. E. M. Joad observed that one already had proof that science could change the body, as in the case of cretins offered thyroid injections and hormones used for one's emotional life. In the future, he mused, "apart altogether from the prospects of determining the sex of our children, we should by gland manipulation, within the next hundred years, make ourselves choleric or timid, strongly or weakly sexed, at will."[70] Not all commentators were as enthusiastic. In *Icarus; or, The Future of Science,* the philosopher Bertrand Russell, gloomily predicted that in years to come the state might employ hormone injections to control the emotions of its citizens. Such powers would all too likely be exploited by the elite. "Assuming an oligarchic organization of society, the State could give to the children of holders of power the disposition required for command, and to the children of the proletariat the disposition required for obedience." And even without hormones the state would learn to use physiology to control the emotions of its citizens. "The chief business of elementary education will be to produce the desired disposition, no longer by punishment or moral precept, but by the far surer method of injection or diet."[71] Russell did not deny that science was making amazing advances. What he questioned was the uses to which science was turned.

Russell was right to be concerned. As soon as hormones appeared, public commentators debated the morality of using them to reform the deviant and discipline the incorrigible. In 1922, *The Times* reported that "certain glands" of an executed Sing Sing prisoner had been transplanted into a habitual criminal.[72] The British did not go as far as the Americans in hormone experiments on the criminal and the subnormal.[73] Nevertheless such approaches did have their adherents. Hormones offered a form of control, a "glandular eugenics'" to fix the unfit.[74] Textbooks asserted that misfits posed a hormonal problem, their endocrine disorders leading to perversions and criminality. Therefore in addition to looking at the possible environmental causes of crime, endocrine investigations should be made. J. B. S. Haldane foresaw glandular substances being exploited for both individual and social reasons. "We already know however that many of our spiritual faculties can only be manifested if certain glands, notably the thyroid and sex-glands, are functioning properly, and that very minute changes in such glands affect the character greatly. As our knowledge of this subject increases we may be able, for example, to control our passions by some more direct method than fasting and flagellation, to stimulate our imagination by some reagent with less after-effects than alcohol, to deal with perverted instincts by physiology rather than prison."[75] In *Cain; or, The Future of Crime* (1928), George Godwin argued that the time would come when the doctor and psychologist would deal with the physiology of the prisoner, not with his or her crime.[76]

Some, such as Bertrand Russell, were alarmed by the prospect of giving such new powers of coercion to the state.[77] Philip Gibbs, in his successful *The Day after To-Morrow,* cited Russell in predicting that the enthusiasm for using hormones to cure deviants and remedy criminals would lead judges to subject the convicted to the mercies of chemistry, surgery, and mental suggestion.[78] In *The Unpleasantness at the Bellona Club* (1928), the mystery writer Dorothy L. Sayers satirized such undertakings. She had one character stupidly say of young criminals, "what they really needed was a little bit of rabbit-gland or something to make them just good as gold." The educated now knew it was possible "to make everybody good by glands."[79] And if liberals like Gibbs were alarmed by the medicalizing of crime, there were on the other side conservatives who bemoaned the idea of felons being treated as pampered patients. In a letter to the editor entitled "Crime and Glands," Sir Robert Armstrong-Jones, physician and psychiatrist, while asserting that he wanted more, not less fear instilled in youth, attacked the reformers' idea that crime was due to glandular deficiencies.[80]

Similar debates raged over the treatment of mental defectives, whose numbers, it was believed, were increasing. The common eugenic argument was that for the Western world to flourish the fertility of the "unfit" would have to be curbed and that of the "fit" rewarded.[81] Sybil Neville-Rolfe, in *Why Marry?,* declared that marriage had to be considered a "privilege" and medical exams were needed so that the public could know "the hereditary qualities of the stocks that are to contribute to the new individual."[82] Eugenicists campaigned for laws that would permit the sterilization of feebleminded. Endocrinologists responded by arguing that a good portion of mental defectiveness was due to hormonal imbalances that could now be treated by doctors.[83] Courses were arranged for magistrates on mental deficiency in which the role of "monkey glands" and internal secretions was noted. Yet British psychiatrists and judges were reluctant to concede terrain to doctors. A typical rejoinder was made by Dr. J. G. Porter Philips of Bethlem Royal Hospital. He was willing to accept that secretions brought on by fear might "poison" an individual, but he could still not accept the notion of "psychical determinism."[84]

Endocrinology was also exploited by eugenicists to shore up notions of racial difference. In talking about "racial endocrine reactions," endocrinologists appeared to lend credence to the racists' claims of essential race differences.[85] Berman declared racial superiority was due to glands.[86] "The point I wish to make," argued Dr. Frank Walker of Toronto, "is that the endocrine glands vary in their activity in different races . . . and that racial crosses are more likely than not to produce instability in the secretions of these glands."[87] The eminent anthropologist Sir Arthur Keith agreed that hormones differentiated races.[88] One commentator asserted that black men like Haile Selassie were adrenal deficient while a self-confident white man like Mussolini obviously had a powerful anterior pituitary.[89] In a *Eugenics Review* article entitled "Race Crossing and Glands," Jon Alfred Mjøen warned that in race crossings the resulting glandular disturbances led to diseases such as diabetes and moral and mental instability that fostered crime. Contrariwise, the closer the parental stocks, the greater the glandular harmony. He concluded his account with the hollow reassurance that he was not a racist. "We shall not persecute other races, but only safeguard our own."[90]

The Sixth International Congress of Surgery held in London in 1923, at which Voronoff and Banting spoke, first heard the opening remarks of Neville Chamberlain, member of the Eugenics Society and the minister of health. "The State was like a hive of bees," he declared. "The drones must never be allowed to outnumber the workers. Anyone who by reason of bodily infirmity was not able to make his full contribution to the service of the community was a drone, a parasite, a burden instead of a source of wealth and strength. Surgery was daily converting drones into workers, and thereby placing the community in its debt."[91] How did the discovery of the internal secretions affect discussions of class? The rejuvenators' attempts at curing impotency can be fruitfully viewed as a form of "positive" eugenics. They unabashedly argued that their efforts were aimed at restoring the vigor and potency of fit, white businessmen and professionals. In the United States, Harry Benjamin wished that all the world's "aging intellectual, political and industrial leaders would be 'Steinached.'" In Germany, Kammerer seemed to suggest that the operation would ultimately open a "path to the Superman." In Britain, Norman Haire argued that testicles should be harvested from hospitals and prisons to "reinvigorate senescent persons of value to the community."[92] The notion that the science might arm one class against another emerged even before the word "hormone" was coined. In *The Food of the Gods* (1904), H. G. Wells produced a sort of genetically modified seed story with a pro-science spin. When children consume a growth substance, conservatives are appalled by the appearance of the resulting "proletariat of hungry giants." Fortunately the miraculous substance spreads throughout the world and all eventually benefit.[93] More pessimistic was Philip Gibbs's *The Day after To-Morrow*. He predicted that the works of people like Steinach and Crew would lead in the future to "a system of state-directed Eugenics or chemical treatment" that would produce workers who would "love their life of toil in the underworld."[94] Given such musings, it is easy to see why Aldous Huxley in *Brave New World* could imagine a society in which doctors tinkered with test-tube babies to produce distinct biological classes. In C. P. Snow's novel *New Lives for Old* (1933), a rejuvenation hormone proves a disaster since only the wealthy can afford it. It exacerbates class hatreds in allowing well-off seniors to pursue sexual adventures previously restricted to the young.[95] As late as the 1950s, Charles Galton Darwin echoed interwar notions that hormones might be used to produce contented celibate workers in an antlike colony. "For example, there might be a drug, which, without other harmful effects, removed the urgency of sexual desire, and so reproduced in humanity the status of workers in a beehive."[96] Given British society's preoccupation with class, it was inevitable that commentators forecast that hormones would be turned to purposes of social control.

However, the most important effect of the early research on the internal secretions was in reformulating discussions of sex and gender. From the moment that Berthold announced in 1849 that testicular implants could turn capons back into roosters, gender came to be regarded as chemically based.[97] The discovery of hormones thus overturned old understandings of the origins of sexual behavior and sexual characteristics. It revolutionized the way in which sexual dysfunctions could be treated. And for staid medical professionals uncomfortable with discussing sexual matters so often associated with quackery, they appeared to legitimize the scientific study of the male reproductive system.[98] Aside from having a placebo effect, we know that none of these sex gland therapies actually worked, but that is not terribly important. They are of interest in that they reveal the ways in which the cultural preoccupations of the age framed the new approaches to sex and gender. Though the investigation of the internal secretions appeared to reconfigure understandings of the workings of the body, they were made part of gender politics as doctors used them to shore up traditional notions of masculinity and femininity.

Given that it was initially assumed that there were "male" and "female" sex hormones, endocrinology's first effect was to rationalize the naturalness of gender difference and heterosexuality. It explained the healthiness of opposite-sex attractions and the sickness of same-sex passions. Hormone research, in revealing how fluid the boundaries were between the sexes, potentially destabilized notions of a clear male/female divide, yet the rejuvenators were clearly intent on maintaining—not eroding—traditional gender boundaries. The sort of gender expectations that doctors brought to their clinics, researchers brought to their labs. Steinach, for example, looked for ardor and aggressiveness in rats as signs of masculinity, and flirtatiousness as a sign of femininity. He took as a given that sex was reflected by behavior and character. Males were supposed to be strong and females gentle. Likewise when researchers turned to human subjects they read off a man's secondary sexual characteristics—voice, beard, muscles, and assertiveness—as indicators of his potency. And they viewed as problematic any signs of "manliness" in a woman. Hormones were needed for health and their lack explained such abnormalities. As a 1924 text asserted, "normal eroticization" depended on these secretions.[99] The physician's duty or desire to render a patient "normal" in turn justified increased medical interventions in the body.

Such explanations were especially helpful in times of change, and gender relations were clearly shifting in the early twentieth century. Interest in sexual rejuvenation was to be expected in an age worried by the decline of the birth rate, the emancipation of women, and reports of flagging male potency. Sexual excesses, venereal disease, and the use of contraceptives were all cited by doctors as leading to impotence.[100] So too was the trauma of the First World War.[101] But civilization itself was also regarded by many as undermining manhood. With the rise of feminism and women's penetration of traditional male spheres came fears that sex differences were being obliterated. In such a context, men's turning to medical means to assure or augment their virility was understandable. In choosing the names for their drugs—"Viriline" and "Testrones" for men and "Fertiline" and "Overones" for women—the commercial firms presented themselves as defenders of heterosexuality. Even progressive doctors viewed the "emancipated" woman as posing a challenge. "The male must still play the dominant part," declared Kenneth Walker, "and if he has sunk to second place in the home he is unlikely to be successful in the sphere of love."[102] Walker gave up providing testicular grafts but believed that hormones treatments did help in improving men's general health.

Much of the explanation for the rejuvenation craze stemmed from the early twentieth-century attack on aging. In his autobiography, Osbert Sitwell playfully described the early twentieth century as "a Paradise where each man and woman, even if no longer born with an immortal soul, could by means of such devices as false teeth and monkey glands have conferred upon them a sort of animal or mechanic immortality of this world."[103] In 1903, Elie Metchnikoff coined the term "gerontology," thereby launching the notion of aging as a medical problem—something that could be tinkered with. In arguing that many men were old before they needed to be, Steinach and Voronoff clearly represented a similar mindset. In the 1920s, doctors increasingly referred to the "male climacteric," which they likened to the female menopause. A man's critical age had more diffuse symptoms than the woman's and occurred later, many doctors placing it the man's fifties.[104]

On the one hand sexual potency was taken as a key indicator of male well-being, and on the other hand retention of sexual vigor was said to delay senescence. Men of the 1920s were exhorted by the media to embrace a new model of masculinity represented by tanned, athletic youths. Like women, men were now supposed to have sex appeal.[105] At the same time, it was still believed that the business life of the man was as important to him as the woman's sex life was to her. Given the social and political disruptions of the post–World War One years, it is no surprise that some men sought out new means of rejuvenation. They found themselves under pressure to keep up with modern women and to counter the challenge of younger business rivals.

The idea of extending youth provided a plotline for a number of English novelists. An extended treatment of the theme is found in John Gloag, *Winter's Youth* (1934). The action takes place in 1960s London, where Dr. Oliver Nordelph is moving beyond Steinach and Voronoff's work by bringing about rejuvenation through applying ultraviolet radiation to the whole nervous system. When the doctor is killed, his notes fall into the hands of Lord Privilege, who immediately recognizes that such a therapy would appeal to the over-forty voter. His party accordingly wins the next election under the slogan "Life-right." Before too long, youth and the working class are enraged to learn that only the rich have access to the process, whereas the eugenically minded British Medical Association insists that it control the service to "weed out the present generation." The orgies indulged in by the political elite scandalize even a voluptuary like Lord Privilege. More seriously, the politicians are not only sexually rejuvenated, they become so combative as to threaten to plunge the country into war. Youth ultimately rebels against both science and the government, but the novel ends with politicians still casting about for ways to use medical therapies for electoral advantage.[106]

Futurists were divided on the value of prolonging life. Lord Birkenhead optimistically believed that in the twenty-first century hormone therapies would extend longevity: "Rejuvenation will be an ordinary and well recognized matter of a few injections at appropriate intervals. The desire to keep old age at bay has ever been one of the dreams of humanity; at last we can predict that it will be achieved."[107] As he made clear in *Brave New World,* Aldous Huxley was more of a pessimist. "Now—such is progress—the old men work, the old men copulate, the old men have no time, no leisure from pleasure, not a moment to sit down and think."[108] He went on to devote an entire novel to the subject in *After Many a Summer* (1939), the plot revolving around the rejuvenation therapies used to keep alive an American millionaire. The sixty-year-old Jo Stoyte is provided with shots of "Testosterons" by a doctor who declares that Brown-Séquard and Voronoff, in pursuing organotherapy, went off on the wrong track. Warnings against any attempts at rejuvenation are offered by a Mr. Propter, who points out the danger that longevity will upset glandular equilibrium and result in degeneration. Indeed the novel ends with the discovery that the one character who for centuries successfully avoided death has evolved into a filthy ape.[109]

How did the discovery of hormones influence the discussion of femininity? Researchers boasted of their abilities to sexually reinvigorate elderly men but showed initially far less enthusiasm for treating older women.[110] Men, doctors assumed, could be sexually active far later in life than women. Nevertheless Steinach and Voronoff declared that that the rejuvenation of women was possible. Haldane echoed their claim that women's old age could be postponed by synthesized ovary gland substances.[111] Lecturing at Cambridge in 1928, Voronoff showed slides of a rejuvenated seventy-six-year-old man and asserted that same results could be obtained by grafting the thyroid, pituitary, and ovarian glands into a woman.[112]

As Chandak Sengoopta has pointed out, the early endocrinologists offered therapies that would serve males and females in different though culturally appropriate ways. Rejuvenation promised men sexual potency and strength (particularly manifested in business success); it offered women primarily a return of good looks.[113] Accordingly, fictional accounts of reactivated women stressed their heroines' sexual restraint. In *The Young Diana* (1918), Marie Corelli, the best-selling novelist, told the tale of a mature woman who runs away from her controlling parents to Geneva, where a scientist gives her a revivifying drug. Years later, more beautiful and youthful than ever, she returns to London. Having enjoyed the delicious revenge of dazzling the man who once jilted her, she retires to a happy celibate life in Paris.[114] A better-known novel depicting female rejuvenation was *Black Oxen* (1923) by the American writer Gertrude Atherton. She herself had heard about the Steinach operation and contacted Harry Benjamin, Steinach's leading disciple in America.[115] He suggested diathermy therapy and employed x-rays that supposedly stimulated her hormones. Before the operation, she was plagued by "mental sterility," but after it she declared that she was energized. "I almost flung myself at my desk. I wrote steadily for four hours." The 110,000 words she produced in the following five months resulted in *Black Oxen*.[116] The book's main male character is infatuated with a glamorous foreign woman with a mysterious past. He is stunned to discover that she is actually fifty-eight but because of having had the Steinach operation looks thirty. Though she promises to marry the passionate young man, she eventually drops him to respond to her higher political calling—rushing to the defense of Austria.[117] In such literary accounts, rejuvenated women were portrayed as desirable and vibrant—not as sexually demanding. They retained their femininity; indeed they enhanced it. They never lost sexual self-control or became aggressive.

Some were still alarmed by the thought that hormones could blur the male/female boundary. In America, the science-fiction writer David H. Keller went so far as to envisage women conspiring to use a male gonadal solution to take over the world. In his feverish imagination, women as a result became masculinized, the female baby sex ratio was increased, and scientists began work on parthenogenesis that would make males completely superfluous.[118] In Britain, the feminist perspective on the employment of hormones advanced by Vera Brittain in *Halcyon; or, The Future of Monogamy* was far more reassuring. Thanks to Steinach's pioneering work, she predicted, future women would age more slowly. By 1973, a serum would be given at menopause that "restored sexual activity of women for a further period of fifteen to twenty years."[119] At the same time, men with excessive sexual desire would be calmed by a glandular operation. A wife could postpone childbearing until she was in her sixties, and since men and women could now synchronize their aging process, their marriages would be further protected.

Women's simple attempts to prolong their youth had their critics. In a study of menopause, Dr. Mary Scharlieb declared that attempts to alter the natural life course inevitably resulted in a variety of physical and psychological complaints.

> A little recognized cause of nervous troubles in middle life and during the climacteric is probably to be found in the unnatural married life that has been gradually gaining ground among us for the last half century. From the seventies onward there has been a growing conviction that it was expensive and undesirable to have a large family. . . . It would appear probable that women who refuse to have their normal quota of children, and who from any motive, whether bad or good, practice prevention, lay up for themselves nervous irritability, discontent, and a depressing sense of failure that are all too apt to aggravate the natural difficulties of the change of life.[120]

One symptom Scharlieb targeted was some women's "quasi-insane desire to prolong sexual life, or to bear children. It is extremely pathetic to find women well on to fifty years of age who are apparently as keen on sexual enjoyment as a bride might be."[121]

M. E. Mitchell's turgid novel *Yet in My Flesh* (1933) carried the same warning against tampering with nature. The central character is a scientist, working on the rejuvenation of rats, who decides to experiment on his wife Molly, a woman who is approaching "the Change." He flatly declares that once passed their fertile years women are "not alive." Luckily they are more easily rejuvenated than men. Molly is eager to try, but the various experiments are plagued by failures. The scientist is disgusted to discover his rejuvenated rats are given to copulating in an "unnatural manner." Molly becomes promiscuous as well. Though the needlessly complicated plot makes it unclear whether the experiment on the woman entails an actual swapping of glands, the author does end the novel by heavily underlining his moral opposition to such procedures. Molly dies due either to a botched abortion or to the injections that have caused a septic womb.[122]

Researchers claimed that hormones would make women better wives and mothers. As Susan Squier has noted, female rejuvenation was justified not just because it prevented aging, but because it promised to "enhance 'normal' familial ties, social relations, and gender roles."[123] In a modernizing world, maintaining one's femininity was regarded by the anxious as increasingly difficult. According to a popular work on rejuvenation, as women become more independent they manifested more cases of vaginismus and lesbianism.[124] In the twentieth century, the healthy married woman was supposed to welcome her husband's embrace. Her failure to do so could now be attributed to a hormonal imbalance. "A further outcome of failing ovarian function is sexual frigidity," declared Steinach. "Apart from these cases of purely psychic origin which are a defence reaction against conscious or unconscious fears, the root may frequently be found in hormone deficiency. If this happens to be the case, then the administration of hormone will render good service, which is so much more desirable, as frigidity often reacts adversely on the psyche, causing feelings of inferiority and endangering marital happiness."[125] A sex manual declared that it was a commonplace that the frigidity of women declined with their emancipation but the sterility of men and women increased with civilization.[126] Fortunately, even the fertility of the mature woman could be reinvigorated by ovarian implantation, claimed Norman Haire. "The senile female becomes more vigorous, shows renewed sexual desire, exerts a renewed attraction over the male, and after a long period of sterility is once more capable of becoming pregnant and producing offspring."[127] Even the ability to give birth to healthy children was now viewed by some as dependent on a woman's internal secretions.[128]

Commercial firms exploited the idea that the purpose of hormones was gendered. By the use of hormones, they declared, women would look better, men would act better. "Normal womanhood" was the goal of female therapies—that is, marriage and maternity. For the first time in history, a generation of women were told they might need to adopt hormonal therapies to keep their sex appeal. Being able to improve the body threatened to shade over into being obliged to improve it.[129] There was certainly a demand for such services. In the late 1920s, Anne Elizabeth Walton of the Radiant Health Institute, Wigmore Street, charged elderly wealthy women a hundred pounds for three months of daily relaxation exercises but found she could demand five hundred pounds for rejuvenation therapies.[130] In addition to employing gland extracts, other practitioners held that frigidity could be cured and the menses restored by the employment of iodine, x-rays, arsenic, ultraviolet rays, radium, and mud baths.[131] And finally newspapers added to their cosmetic advertisements purportedly new scientific products such as "Leichner Hormone Cream," guaranteed to rejuvenate the skin.[132]

Although the rejuvenators primarily focused on men, the most dramatic advances in sexual endocrinology in the interwar years were made in understanding the role of female hormones. One of their more practical though often overlooked uses was in pregnancy testing. In 1927, the German scientists Selmar Aschheim and Bernhard Zondek discovered that if the urine from a pregnant woman was injected into a mouse, it would go into heat. Thanks to this revelation, joked R. W. Johnstone, a Scottish professor of midwifery, rodents were raised to the "rank of obstetrical consultants." We now know that the urine of a pregnant woman contains a hormone—the human chorionic gonadotropin (HCG)—secreted by the placenta.[133] F. A. E. Crew and Bertold Paul Wiesner followed up on this work to establish in Edinburgh a Pregnancy Diagnosis Station (using the urine test) that served the physicians and hospitals of Britain and Northern Ireland. The ten thousand tests a year that it was carrying out by 1939 determined more than just pregnancy. A hormonal count could indicate danger of miscarriage or fetal abnormality and the differences between pregnancy, abnormal emotional states, and menopause.[134] "However imperfect and expensive," concludes Ann Oakley, "the A-Z test launched the modern era in which obstetricians would eventually be able to claim a knowledge superior to that possessed by the owners of wombs themselves, as to the presence of a guest, invited or uninvited, within."[135]

Aschheim and Zondek's research also resulted in scientists focusing on urine analysis in their search for the female hormone. As a result, in 1929 both Adolf Butenandt in Germany and Edward A. Doisy and Edgar Allen in the United States declared that their teams of researchers had isolated estrone.[136] By 1931, pharmaceutical corporations were marketing it. In addition to the estrogens that elicited the development of the secondary sexual characteristics, scientists also isolated progestrone needed for gestation. To sort out the rush of research results, the First International Conference of Standardization of Sex Hormones was held in London in 1932. Having isolated the sex hormones, the question was what to with them. "Sex hormones," Nelly Oudshoorn notes, "may best be portrayed as drugs looking for diseases."[137] If hormone treatments for women took off faster, it was not because they actually needed them but because women's complaints were medicalized sooner. The hospitals and clinics already dealing with women's reproduction provided scientists with access to a compliant clientele. There was no similar network of male clinics that could be exploited. In employing such modern drugs, gynecologists raised their professional profile. However, progress in Britain was slower than in the United States, where endocrinology and gynecology "practically merged" in 1930s and hormonal treatments were prescribed for everything from girls' late puberty to the "unhappiness" of the menopausal.[138]

In 1922, T. S. Eliot attributed his wife Vivienne's mental instability in part to her having been abandoned by her lover Bertrand Russell, but Eliot told the American poet Ezra Pound that he believed that her problems were essentially due to her glands. He was impressed by the hormonal diagnosis of Lancelot Hogben who prescribed "Ovarian Opocaps."[139] From the beginning the question was posed if the knowledge of the role purportedly played by the internal secretions would be liberating or controlling. That hormone manipulation should serve social purposes was made clear in the pioneering work of Bertold Paul Wiesner (1901–1972), PhD, DSc, an Austrian-born sex physiologist.[140] As early as 1926, he presented a paper "Concerning Germ-Gland Functions in Infancy" at the First International Congress for Sex Research organized in Berlin by Albert Moll.[141] In

1928, Wiesner joined F. A. E. Crew's Department of Research in Animal Breeding (in 1930 renamed the Institute of Animal Genetics) at the University of Edinburgh, as head of the sex physiology section, where he worked on gonadotrophic hormones. As early as 1925, he had investigated the prevention and interruption of pregnancy by physiological means, or as a journalist reported "natural-artificial birth control, that is, the control of birth at will but without recourse to manufactured means."[142] The female endocrine system was, he declared, more complicated than the male. Hormones could induce or prevent ovulation and a balance of hormones was needed for a normal pregnancy, but they affected more than reproduction.[143] Wiesner's funding came from the Rockefeller Foundation and the Eugenics Society, two organizations very much interested in developing a foolproof contraceptive.[144] Thus in the 1920s and 1930s, some of the basic research was carried out that in the post–World War Two era resulted in millions of healthy women being prescribed the birth control pill and hormone replacement therapies.

Given that scientists were claiming that almost every aspect of a woman's sexual life could be remade by hormonal manipulation, it comes as no surprise that some would envisage actually "making" a woman. The best example is Edward Heron-Allen's novel *The Cheetah Girl*. Privately published in 1923 under the pseudonym of "Christopher Blayre," it was undoubtedly the most sensational of all the interwar novels sparked by reports of the rejuvenators' work on hormones, reproduction, and cross-species transplants. The pseudonymous author carefully distances himself from his pornographic tale by claiming simply to be the editor of a manuscript given to him by a lawyer who obtained it in turn from the unnamed narrator, a professor of physiology. The professor begins his story by making the startling assertion that he is going to kill his wife, and he then proceeds to explain why. He says he was trained in medicine by one Paul Barrowdale, a medical scientist, who, upon his death bequested to him his child Uniqua, a tawny-skinned, seductively lithe creature, whose rough tongue and sharp teeth make her kisses intoxicating. The adolescent is especially passionate when in heat and, despite her mother's desperate though unexplained opposition, seduces and marries the narrator. When she becomes pregnant, her mother commits suicide. Why? The narrator finally learns the truth when he reads the memoir left by Barrowdale. Fascinated by experiments in "artificial fertilization" by such scientists as Jacques Loeb and E. Ray Lankester, Barrowdale relates that he was especially impressed by reading that barrenness could be cured by injections that could overcome a couple's hostile toxins. He muses that difficulties with blood transfusions might be similarly dealt with, but decides to experiment himself in cross-species fertilization.[145] He employs a prostitute, "Ménagerie Sal," the offspring of a homosexual and a lesbian whose lascar lover has already subjected her to intercourse with a variety of beasts. Barrowdale uses inoculations to render her and a cheetah sexually compatible. Six months later, Uniqua is born.[146] Barrowdale realizes too late the results of his successful attempt at hybridization. The beautiful Uniqua is sexually mature at thirteen and in heat every three months. If she is not prevented from reproducing, the race is doomed. The narrator, now terrified by his reading of this memoir and having failed in convincing the enraged Uniqua of the need to abort, concludes his account by surreptitiously preparing to inject her with a murderous dose of aconitine. We know that he did so as the editor's concluding remark is that the professor and his wife died twenty-five years earlier.[147]

Heron-Allen's torrid tale had a miniscule readership, and no claim can be made that it influenced public opinion. Nor was it particularly original, being a variant of the longstanding male dream of manufacturing a compliant sex partner. It is obvious, however, that in referring to artificial fertilization, cross-species breeding, and hormonal injections it was reflecting interwar interests in how women could be chemically made or remade. In fact, the world heard that a man had been made into a woman in 1930 when the Danish painter Einar Wegener underwent a sex change operation and emerged as Lili Elbe.[148] Wegener had denied being a homosexual and believed that his body contained a woman's sex organs. His sympathetic doctor cited Steinach to argue that Wegener's rudimentary ovaries might be activated if fresh ovaries were implanted. He was subjected to a series of operations that were apparently successful, but Lili died soon after of complications. In his introduction to an account of the affair, Norman Haire reported that Steinach did have success in carrying out such operations on lower animals. Having observed in Germany a male who failed in his attempts by hormone injections and operations to change his sex, Haire believed that too little was as yet known about such procedures and it was best to turn to psychological treatments to "cure" or help to adjust such patients. Nevertheless the moral of both *The Cheetah Girl* and the Lili Elbe story appeared to be that women could be created.

Interwar commentators found the opposite scenario—making a man out of a woman—more difficult to envisage. It is true that in 1936 under the headline "Woman Changed to Man Faces New Life," the *Daily Herald* reported that Mary Weston, a female champion javelin thrower and shot-putter, had been recently operated on by Dr. Lennox Broster of London's Charing Cross Hospital and emerged a man. Though *Time Magazine* also declared it a "Change of Sex," Broster insisted that the problem simply resulted from an initial mistake in gender assignment. "Mr. Mark Weston [the patient's new name], who was always brought up as a female, is male, and should continue life as such."[149] Stressing that Weston married a woman soon after his operation, the press underscored the fact that he was not a homosexual.[150]

How did the emergence of hormones influence the discussion of homosexuality? Doctors who asserted their ability to rejuvenate femininity and masculinity tended to regard as abnormal any blurring of gender. Moreover they claimed that the same operations that restored vigor in normal males could kindle heterosexual drives in the abnormal. Voronoff stated that he had successes in four of the seven cases in which gland transplants were employed to remedy sexual inversion. Steinach agreed that homosexuality was due to a hormonal deficiency. After the urologist Robert Lichtenstern did a testicular graft for an accident victim, Steinach asked him to provide a homosexual with the testicle of a "healthy man." Homosexuals were obviously not healthy. Steinach asserted that the testicles of five of his homosexual patients in fact had cells resembling those found in the ovaries. Though the Spanish scientist Gregorio Marañón believed Steinach's observations were faulty, he still felt his colleague was on the right track. Marañón advised two homosexuals to undergo the Voronoff treatment. In one case, it appeared to work, though Marañón concluded that suggestion probably played a role.[151]

Louis Berman claimed that homosexuality was due to thymus problems, and he went on to explain that if Oscar Wilde were alive in the 1920s, he would soon get help.[152] Others agreed. "Those with a persistent thymus are subject to all shocks and sex perversions. Homosexuality is not psychic. It is usually thymic. . . . The pituitary holds certain sway that is very vital. Its growth depends upon the size of the sella turcica. If it is too small there is not only a poor physical but mental development as well. There is a difficulty in learning or retaining anything, degeneracy of the moral and sexual type, epilepsy and sex perversions."[153] Moreover the fact that "female" hormones were found in male urine suggested a fluidity of boundaries and for some seemed to provide a physiological explanation of homosexuality.[154] In 1923, Julian Huxley declared that the enlightened now regarded the hormones as central in determining sexual temperament.[155] Citing Steinach's belief that homosexuality could be cured by a testes transplant, Huxley judged that the process was not yet proven but in theory made sense. In animals, freemartins had both male and female characteristics. In humans, male abnormals looked feminine due to delayed activation of the testis. So it was not their mind but their physiology that was at fault.[156] The sellers of hormonal quack remedies quickly exploited these views with such claims as "Viriline tends to prevent the appearance of sexual perversities."[157]

Most conservatives regarded homosexuality as a failure of the will, but some would accept that hormone imbalances led to sexual perversions. According to Sybil Neville-Rolfe, the sex glands in civilized life were artificially stimulated resulting in an abnormality that, she declared, "is socially and racially very undesirable."[158] Proponents of eugenics like Neville-Rolfe frequently suggested that "perverts" be included as potential subjects of either castration or sterilization. A number of the medical scientists interested in rejuvenation techniques were involved in such undertakings, though in giving gland transplants and hormonal injections they saw themselves "curing" rather than punishing their patients.

Magnus Hirschfeld, Europe's leading campaigner for the rights of homosexuals, supported such experiments. Since he previously claimed that true males and females were only ideals and the vast majority of individuals found themselves somewhere on the continuum in between, his new stance appeared somewhat contradictory. He took this position because those hostile to homosexuals commonly argued that they simply had made the choice of being perverts. Hirschfeld countered that scientists had established that homosexuality had a physiological basis, it was neither a vice nor a disease but a "developmental error." He asserted that the interstitial cells of homosexuals were distinctly different from those of heterosexuals and elaborated a theory that their testicular secretions led to development of a more "feminine" type of brain. He naturally took great interest in Steinach's work. So, on the one hand Hirschfeld used hypnotism to heighten the desire of heterosexual men suffering from impotence and on the other he sent a handful of patients who wanted to overcome their bisexuality or homosexuality to surgeons like Richard Mühsam. The failure in one case he attributed to the fact that though the patient received a "normal" testicle, it had to share a scrotum with its homosexual counterpart. Arguing that homosexuals had no choice in deciding their sexual orientation made tactical sense for a man seeking an end to the law that declared them criminals, but in going on to argue that homosexuals were physically different from normal males Hirschfeld appeared to be implying that they were in some way sick and so could possibly be cured.[159]

Strikingly enough, some of the doctors who enthusiastically supported offering hormonal therapy to those who wished to be "cured" of their sexual orientation were themselves homosexuals. Norman Haire, for example, publicized the Steinach operation and praised the attempt made by Robert Lichtenstern in 1916 to cure a homosexual by providing him with normal testicle. As Haire told it, within a short time the patient destroyed the will in which he had left all his money to a male friend, developed secondary sex characteristics, flirted with his nurse, and finally married.[160] Kenneth Walker more cautiously noted in 1930 that some doctors thought homosexuality was "the result of disturbances of the endocrine function of the sex gland, or some other organ of internal secretion," yet he maintained that "the surgical treatment of homosexuality cannot be considered a practical measure until much more is known, not only of the cause of homosexuality but also of tissue grafting in general."[161]

By the late 1930s, a researcher like Bertold Wiesner was stressing that sexual abnormalities were not simply due to hormonal disturbances; they had multiple causes.[162] Similarly the American William J. Robinson regarded homosexuality as a "pitiful abnormality," perhaps caused by an endocrine imbalance, yet he concluded that there was no known cure.[163] Such attempts to attribute homosexuality to hormonal imbalances would continue to flourish long after the decline of organotherapy.[164] When gland grafting declined, doctors turned to hormonal injections and surgery. Charing Cross Hospital surgeon Lennox Broster ended his survey of the "surgery of sex" by declaring that though the evidence was still sparse, one had to hope the glands were the cause of sexual abnormalities because they could at least be operated on. By the 1940s, it was accepted by medical scientists like Broster that everyone was born inherently bisexual because each sex produced both male and female hormones. Nevertheless he regarded it as his duty to restore "normality" via surgery to men suffering from "feminism" (wide pelvis, slender limbs, hairless, high voice) and virilized women.[165] A whole range of social factors including shifts in fertility and women's greater participation in public life was, according to Broster, leading to racial degeneration. One of its manifestations was sexual abnormality. "The presence of the adrenogenital syndrome in man is a retrograde movement, since it tends towards the intersex type. Is it stationary, receding, or increasing? If the latter," warned Broster, "then we may be the innocent spectators of an evolutionary process drifting slowly and inevitably into the neuter state."[166]

It is striking how much of the early discussions of endocrinology was framed by the eugenic notion that medical scientists had a duty to improve the race. Most of those who defended hormonal therapies took it as a given that the reproduction of the "fit" should be encouraged while that of the "unfit" should be curbed. They tended to define healthiness as a manifestation of good biochemistry. Misfits, it was assumed, suffered from a hormonal problem. In dealing with them, the doctor's duty was to restore a "natural order."[167] In short, the discovery of hormones had recast eugenics. This argument was forcefully made by the American Louis Berman, in *The Glands Regulating Personality* (1921), who hailed the new science for promising to improve the breeding of the healthy while preventing the matings of the defective.[168] Berman was in turn hailed by Ezra Pound for publicizing a doctrine that, in his estimation, equaled Darwin's in importance and dwarfed Freud's.[169] Though Julian Huxley was disappointed that Berman's shoddy study was full of such bizarre statements as "not utilitarianism . . . is needed but pituitarianism," he conceded that temperament was now recognized as more important to success than intellect, and temperament was produced by biology.[170]

A writer of a 1924 medical text on the internal secretions asserted that it was essential that eugenicists, because of their preoccupation with breeding, have some knowledge of the workings of the ductless glands.[171] The eugenically minded for the most part believed that clear demarcations separating the sexes were the basis of good breeding. Given such a concern, they naturally welcomed those efforts made by endocrinologists to shore up masculinity and femininity. Endocrinologists were drawn to eugenics; what did eugenicists make of hormones? The fact that the *Eugenics Review* kept its British readers up to date on the works of the rejuvenators demonstrates that its readers were not all simple biological determinists and were interested in the ways in which the hormones might intervene.[172] Some eugenicists naturally enough regarded endocrinology as a challenge to the hereditarian argument that biology was destiny. Voronoff claimed that a new vigor enjoyed by the parent could be inherited by their young. If true, the eugenicists acknowledged, such a process would be of the "utmost importance."[173] Some indeed appeared to believe that hormonal treatments might provide a benefit that could be passed on to a subsequent generation as a sort of acquired characteristic.[174] However, most contributors to the *Eugenics Review* swung back to the more orthodox view that the existence and activity of glands were themselves determined by heredity. N. B. Dreyer condemned gland rejuvenation operations as wrong and dangerous. F. A. E. Crew was skeptical of Steinach and Voronoff's work.[175] In the late twenties, Voronoff's work came under attack, with Arthur Walton (a member of a commission investigating Voronoff's animal transplant data) insisting that no claims for rejuvenation had as yet been scientifically proven.[176] By the early 1930s, the *Eugenics Review* seems to have dropped the discussion of rejuvenation.[177] But rejuvenation therapies continued to be noted in manuals linking sex and eugenics.[178] And though rejuvenation was derided, many were still fascinated by the notion doctors could provide hormones to improve human efficiency. It took little imagination to envisage how a perfect eugenic program could be implemented if gland extracts were available to rejuvenate the elderly, speed up workers, and reform the deviant.

Though increasing the sexual vigor of the otherwise healthy and limiting the reproduction of the "inferior" might appear to have little in common, they were regarded by some in the decades between the world wars as two sides of the same coin. Rejuvenation, eugenics, and endocrinology were each in their own ways symptoms of the twentieth-century desire to improve the body. They spoke to the faith so many had in the early twentieth century that medical science could solve every sexual conundrum. It was no accident that the interest in rejuvenation research first crested in the 1920s. Given the huge population losses during World War One and the subsequent Spanish influenza epidemic, any undertaking that promised regeneration of the population would likely win a hearing. Rejuvenation became the rage in the same decade in which Fascists in Italy, Nazis in Germany, and Communists in the Soviet Union all asserted their intentions of creating new men and new societies.[179] But in Britain as well, the notion of science now being able to make humans more efficient, virile, and productive was very much in the air.[180]

Scholars have pointed to the difficulty in drawing a clear line between the "pursuit of health" and the "elimination of unfitness."[181] By the 1920s, this enhancement culture was already in the process of emerging and clearly linked to eugenic preoccupations. Hormones were to be used both to repress the passions of the deviant and to restore the vitality of the normal. The existence of such dialogue between eugenics and endocrinology reminds us of the need to avoid exaggerating the difference between "positive" and "negative" eugenics. Whether it was in encouraging the reproduction of the fit or limiting that of the unfit, defending heterosexuality or "curing" sexual deviancy, the eugenically minded sought to turn endocrinology to the purposes of policing reproduction.

CHAPTER 5

"A spinster and a syringe"

Debating Test-Tube Babies

The play *R.U.R.* ends with the last living human being warning the robots: "life will not proceed from test-tubes."[1] Some in the 1923 audience likely believed otherwise. Hormonal therapies appeared to offer the means by which men and women could be rejuvenated or remade. Moreover the scientifically inclined in the interwar years envisaged the even more radical prospect of the test-tube or ectogenetic baby. We have already noted that J. B. S. Haldane spelled out how human eggs could be harvested and ectogenetic children produced.[2] Haldane's vision inspired a number of writers including J. D. Bernal, Norman Haire, H. J. Muller, Charlotte Haldane, and Vera Brittain. In the famous opening passages of *Brave New World*, Aldous Huxley describes in some detail how women are induced to have their ovaries removed and their eggs collected, and depicts the Central London Hatchery and Conditioning Centre, with its Fertilizing Room, Bottling Room, Social Predestination Room, and Decanting Room. Huxley is, however, curiously silent about the male sperm donors.[3] So too were most of the other commentators who speculated about ectogenesis. Such an oversight is striking given the fact that at the time the media were carrying stories of babies conceived as a result of artificial insemination. The appearance of such children—referred to in the newspaper press as "test-tube babies"—was interpreted by some as the most dramatic demonstration possible of the benefits of reproduction by design; by others of its inordinate ethical costs. Gaining a clear understanding of early British attitudes toward artificial insemination is challenging because many regarded the subject as so inherently unsavory that it should not even be publicly discussed. For that reason, this chapter approaches the issue in a roundabout fashion by first investigating a relatively obscure Canadian court case.

On November 23, 1920, under the headline "Sordid Story Told in Court," the *Toronto Globe* informed its readers that "evidence of such a revolting character" was to be heard in the Orford vs. Orford alimony suit that Mr. Justice Orde had decided that much of the proceedings would be held in camera. The trial that took place in Osgoode Hall in late November and early December of 1920 was indeed one the most sensational Canadian court cases of the twentieth century. On the surface it was a simple suit for alimony, launched by Lillian Orford against her estranged husband Frederick Orford. The real significance of the trial, however, was that it marked the first time in history that a common-law court dealt with the subject of artificial insemination. The obvious question is, why did the issue surface at this particular time and place? Seeking to provide some answers directs our attention in turn to the ways in which commentators in the interwar years attributed enormous social significance to a rarely employed but simple form of medical treatment.

Let's begin with the Orford trial.[4] Toronto readers hankering after sordid details must have been disappointed by the brief portion of the proceedings that the newspapers felt fit to print. It consisted mainly of Lillian Orford's testimony. She told how, in 1910, she had left her parents' home in Weston-Super-Mare, England, to come to Toronto to care for her invalid married sister. In addition, having studied music in England, Lillian continued to do so at the Toronto Conservatory. On April 1, 1913, she was introduced to Frederick Orford, a wealthy businessman of forty-six, about twice her age. Described in the press as a "capitalist and financier" and a "Toronto real estate broker," he purportedly had an annual income of $25,000 and was worth an estimated $500,000.

If Lillian was impressed by his wealth, Frederick was smitten by the accomplishments of the young English woman. The day after their meeting he sent her a piano as a present. According to Lillian, he then began a curiously erratic courtship. Occasionally breaking down in tears, he admitted to be suffering from some obscure mental complaint. This would have put off some women, but, as he was much older, Lillian claimed that she assumed that Frederick could be trusted. Accordingly she was bold enough to accept his invitation to accompany him on a brief visit to New York City where he promised exposure to Christian

Science and the mind cure movement known as "New Thought" would both do them good. Expecting that they would have separate rooms in the New York hotel, she was disconcerted to discover that he had rented a suite. Lillian slept the first night fully clothed. More alarming was her seeing Frederick surreptitiously pour a powder into her lemonade. When she confronted him and threatened to call the hotel staff, he sought to reassure her by downing the drink himself. He immediately became drowsy. While he slept, she rifled through his suitcase and found chloroform, strychnine (often used as an aphrodisiac), and morphine. On awakening, the rueful Frederick begged for forgiveness and asked her to pray with him.

Back in Toronto, Frederick rented rooms on Pembroke Street for Lillian, who in exchange agreed to act as a nurse and companion for his mother. During a visit, he again gave Lillian a suspect drink that she made the mistake of taking. It made her dizzy, and she sought the help of neighbors. Frederick, whom she accused of being responsible for her light-headedness, had her examined by doctors. Soon after he drove her and a nurse into the cottage country of Muskoka where Lillian convalesced for several weeks.

Given such mysterious goings-on, why did Lillian simply not break with Frederick? He appeared to be more in need of a nurse than a wife. In Toronto, he had Lillian apply an ointment for a facial skin complaint he had. He undertook the trip to New York in the hopes of mentally refreshing himself. In his letters read in court, Frederick admitted to having difficulties in keeping his balance, to using sleeping powders, of having been told by Dr. Elliott to stop working for a time, of toying with the idea of going sixty days without talking in order to cure his "nerves."

Lillian's own evidence suggests that they were both playing risky games. Presumably in her eyes, Frederick's wealth compensated for his strange behavior. The latter was capped at the summer's end by his declaring his love and asking her to marry him. She did take his ring and agreed to wed on condition that they visit her parents in England. Even so, she confessed to needing a brandy to go through with the service. On August 26, 1913—less than five months after first meeting—the couple was married in Toronto by the Reverend Anthony Hart. On September 5, they left for England on board the Lusitania. As might have been anticipated, the honeymoon voyage was a disaster. The marriage was not consummated, or at least not in the usual sense of the term. In her statement of claim, Lillian accused her husband of "brutal, immoral and unnatural treatment," implying that he had resorted to some perverted form of intercourse.

Once in England, Lillian went on to visit her parents in Weston-Super-Mare.

Frederick stayed in a nearby hotel. On November 5, 1913, he returned to Toronto. He suggested that Lillian remain in England with her parents. Indeed Lillian testified that he had stated that the marriage had been a mistake and that he envied "a friend of his who liked to see his wife about every six or seven years."[5] When she later wrote about her returning to Canada, Frederick put her off. As it turned out, she was not to be in Toronto again until after the ending of the Great War.

Frederick was apparently not a well man, but more central to this case was the fact that Lillian had her own medical issues. Frederick's idea was that she remain in England to "take treatment."[6] Lillian protested in court that she was now being castigated when all her efforts were undertaken "to fit herself to return to her husband."[7] Her letters from England referred to consultations with doctors and the possibility of an operation. What sort of operation? The press reported that the judge was skeptical that such an "important operation" had ever been carried out, particularly given the fact that the husband had not been told of it. "He was the one man in the world," Mr. Justice Orde declared, "who ought to have been consulted."[8]

On December 24, 1920, the trial concluded, but Mr. Justice Orde reserved his judgment. The presumably puzzled newspaper-reading public had as yet no sense of Frederick's response to Lillian's charges nor of the basis for his counter-argument that she was guilty of adultery. Only those who read Mr. Justice Orde's judgment, made on January 5, 1921, could finally make sense of the case. What the press had not reported and the judge now revealed was that the marriage had not been consummated because of pain that Lillian experienced when intercourse was attempted and her later having been diagnosed as having a retroflexed uterus. In frustration Frederick had fallen back on what she described as "unnatural practices." He denied this but accepted that they both agreed she would remain in England to seek a cure. But when she did return to Canada in December of 1919, he refused to accept her as his wife, and she accordingly began her alimony suit. To counter it, Frederick in April 1920 launched divorce proceedings. In June 1920, his first divorce petition was turned down. Frederick then had investigations made in England. They revealed that in February 1919 Lillian had given birth to a child fathered by one George Edmund Hodgkinson.[9] Frederick's argument was that his wife was an adulteress, and consequently he owed her nothing.

Lillian admitted giving birth to Hodgkinson's child. How then could she persist in insisting that she was not guilty of adultery? Peter White, Lillian's lawyer, advanced a remarkably daring argument. He conceded that Lillian's child was not Orford's, but asserted that by law she had not committed adultery as she had conceived as a result of artificial insemination. Her story was that in July 1917 she met Hodgkinson in London. Later that fall, she saw Dr. Rayner about her gynecological problems.[10] In November 1917, she also consulted the unfortunately named Dr. Stabb.[11] In January 1918, she again saw Hodgkinson and told him that Dr. Rayner said that she needed a special operation but that he would not provide it without her husband's consent. He added that the only other cure would be by bearing a child, which could be done artificially. Lillian told the court that she confided in her good friend Hodgkinson who appeared to know "quite a little about insemination."[12] If she were willing to attempt the procedure, he offered to supply "what was necessary," pay all expenses, and adopt the child. In his flat, an unnamed physician assisted in a first attempt in March and a second in May. On each occasion Lillian was put under an anesthetic, and the semen introduced via a syringe (or at least that was what she said she was told by Hodgkinson). She became pregnant as a result and a child, subsequently baptized "Peter Lee Hodgkinson," was born February 13, 1919.[13]

"I was trying to cure myself for my husband," Lillian insisted, "that was my only excuse." Mr. Justice Orde was not convinced. Indeed he was outraged that she refused to recognize the seriousness of her act. She spoke of a "medical cure" when there was nothing medical about it; of "artificial insemination" when there was nothing "artificial" about it either. In fact, Orde believed that her story was a complete fabrication, that her physical problem had in some way been resolved, and that Lillian had committed adultery with Hodgkinson "in the ordinary way." Nevertheless the judge still felt obliged to go on to deal with the argument that Lillian's counsel advanced that a woman "artificially inseminated" by a man other than her husband had not committed adultery. White's point was that in law adultery always meant the "sexual connection between a man and a woman, of whom one at least was lawfully married to a third person" and without actual intercourse there could be no adultery.[14] Orde disagreed and declared that adultery consisted of "the invasion of the marital rights" of a spouse and the chief right and first purpose of marriage was the bearing of children. Orde reiterated the traditional double-standard view that "Sexual intercourse is adulterous because in the case of the woman it involves the possibility of introducing into the family of the husband a false strain of blood." For Orde, the essence of adultery was not so much the act of intercourse per se as "the voluntary surrender to another person of the reproductive powers of the guilty person."[15] Accordingly Orde concluded that although earlier legal scholars had never envisaged such an act, a woman's recourse to artificial insemination by donor ("the introduction into her body by unusual means of the seed of a man other than her husband") was clearly adultery.[16] Orde found in favor of Frederick Orford, thereby dismissing Lillian's alimony suit. Armed with this judgment Frederick finally obtained his divorce on June 4, 1921.[17] Because of the ways in which Justice Orde expanded the definition of adultery, legal scholars would continue to cite his decision for decades to come.[18]

What is surprising is that although the Orford case marked an important legal benchmark in the discussion of artificial insemination, historians of medicine have so far ignored it. To judge by the best accounts of assisted reproduction such as Naomi Pfeffer's *The Syringe and the Stork*, artificial insemination was not practiced in Britain until the late 1930s. Now, of course, Lillian Orford's story of being artificially impregnated might well have been a tissue of lies. But that is not the point. The question is, from where did she get her ideas? In trying to find an answer, it becomes clear that in the early twentieth century just as the subject of birth control split the British medical world into radical and populist types in favor of such discussions and the professional elite who were opposed, remarkably similar responses were made to the issue of artificial insemination. Two additional points are made by extending our investigation into the 1930s and 1940s. The first, which is hardly surprising, is that a procedure seen by some as a welcomed remedy for the unhappily infertile was long regarded by others as posing a threat to Christian morality, traditional gender relationships, and the respectability of the medical profession. The second is that such debates revealed the extent and tenacity of eugenic notions in Britain. The argument that they were motivated by a desire to improve the race was wielded by both the opponents and defenders of artificial insemination.

Discussions of artificial insemination, of course, predated the twentieth century. A fifth-century Babylonian Talmudic passage refers to the possibility of a virgin conceiving in a bath in which a male left semen, a thirteenth-century rabbinical pronouncement held that a child born as a result of a woman lying on the sheets of a man not her husband was still legitimate (since no intercourse occurred), and a Midrashic legend of the fourteenth century stated that Jeremiah's daughter was made pregnant by bathing in his bathwater.[19] Such notions had a long life. In *Vulgar Errors* (1646), Sir Thomas Browne felt obliged to rebut the report that a woman could be passively impregnated by bathing in a tub in which a man had left semen: "'Tis a new and unseconded way in History to fornicate at a distance, and much offendeth the rules of Physic, which say, there is no generation without joint emission, nor only a virtual but corporal and carnal contraction."[20]

A nineteenth-century American text on medical jurisprudence noted the continued popularity of such ideas. It reported that a woman, after making love to her husband, purportedly impregnated her female partner. More sensationally, it stated that

> Among the cases in the United States Pension Office . . . is a claim for a widow's pension in which the claimant declares that at one of the battles of the late war she was a spectator, and happening to stand behind some soldiers, was wounded in the abdomen by a ball that had previously wounded one of the soldiers in the testicles, from which she became pregnant and was delivered of a boy.[21]

Actual successful attempts at artificial insemination date back as far as the eighteenth century.[22] The Italian abbé Lazzaro Spallanzani produce a treatise on "artificial fertilization" in 1779, having inseminated frogs and a Barbet bitch. One might note that Spallanzani supported the ovist idea that the mammalian egg was simply stimulated by semen and that the "animalcules" or spermatozoa observed in it, being mere parasites, played no active role.[23] In 1776, the Scottish surgeon John Hunter supervised the first successful attempt at human artificial insemination when he instructed a linen draper, afflicted with hypospadias, on how to use a warm syringe to impregnate his wife. The attempt at artificial insemination by husband (or AIH) worked and pregnancy ensued, but because of Hunter's fear of the criticism of moralists his brother-in-law only reported it posthumously twenty-three years later in the *Philosophical Transactions* of the Royal Society.[24]

In the English-speaking world, the American gynecologist Dr. James Marion Sims (1813–1883) is often cited as the first doctor to publish an account of artificial insemination.[25] He certainly precipitated a lively debate by his 1866 book, but most nineteenth-century discussions of artificial insemination actually took place in France. As early as 1803, the French were following up on Spallanzani's work.[26] Dr. Girault began in 1838 to provide a series of cases that he reported on in 1861 and 1869. Making his injections by blowing semen through a hollow sound, usually two days after menstruation had ceased, he asserted that ten of twelve attempts ended in pregnancies.[27] According to Gérard, the sperm was collected either by "la méthode Française" (masturbation) or "la méthode Americaine" (syringed from the vaginal basin).[28] What appears to be the first reference in a Canadian publication to artificial insemination took place in 1870 when the *Canada Lancet* reprinted an account of Girault's paper.[29] The issue was debated by both the French legal and medical professions.[30] In 1883, as a result of the "Affair Lajartre," a Bordeaux tribunal declared that artificial insemination, even by a husband, was a practice "contrary to the natural law and [a practice] which could constitute a veritable social danger."[31] Nevertheless the Société de Médecine Légale formally defended the legitimacy of the operation.[32]

In an overview of the subject in 1889, Jules Gautier made two points worth underlining. The first was that the ancients thought the woman's passions were needed for conception to occur but doctors now regarded her orgasm as superfluous. The second was that the injections usually took place just before or after the menses when doctors believed conception was most likely to occur.[33] Gautier was puzzled, however, by the fertility of the Jews, since they avoided these very days because of the taboo against intercourse when the woman was considered unclean.[34] Given the fact that Ogino and Knaus's correct determination of the ovulation cycle was only popularized in the 1930s, it is indeed hard to credit many nineteenth-century claims of success.[35]

Nevertheless the interest in artificial insemination was so wide spread in France that by the 1880s it became a key theme in a number of novels.[36] With the lowest birth rate in Europe, it is hardly surprising that the French would produce accounts of what Michael Finn refers to as "problematized reproduction."[37] At the turn of the century, one could even find French cartoonists satirizing the interest in "la fécondation artificielle."[38] Passages describing artificial reproduction by syringe became a staple of early twentieth-century French marriage manuals.[39]

The social impact of artificial insemination practices was speculated on from the mid-nineteenth century. Auguste Comte suggested that some means of artificial insemination be invented so that women could remain unsullied by the unruly passion of sexuality.[40] In the 1890s, the conservative social theorist Vacher de Lapouge envisaged using artificial insemination to eugenically perfect humanity "using a very small number of males of absolute perfection . . . to inseminate all the females worthy of perpetuating the race." As one healthy man could produce two hundred thousand births, the world could be perfected in a few centuries. By "la fécondation artificielle," in Lapouge's view, scientific reproduction replaced the bestial, love, lust and reproduction were split, and Minerva thus usurped Eros.[41]

What of the morality of the practice? Although its defenders argued that artificial insemination served a moral good in assisting rather than countering nature's intent, the process nevertheless required the man to produce semen, and masturbation, in the eyes of the Catholic Church, was a sinful act. But since the goal was to allow a couple's marriage to serve its first purpose—the procreation of children—would the church countenance it?[42] In 1887, the question was put to the Sacred Congregation. The Holy Office of the Catholic Church curtly responded in March 1897 that the process was illicit or "non licere."[43] Nevertheless the topic continued to be debated by Catholic writers well on into the twentieth century.[44]

Italians also discussed the issue. Paolo Mantegazza is often cited as the first doctor to refer (in 1887) to artificial insemination by donor. He also mused about the storage of sperm. "It might even be that a husband who has died on a battle-field can fecundate his own wife after he has been reduced to a corpse and produce legitimate children after his death."[45] Americans discussed the issue. The first reported case of artificial insemination by donor (or AID) in the United States was reported in the *Medical World* in 1909.[46] Robert Latou Dickinson, considered a medical maverick because of his outspoken support of birth control, also described the employment of donor sperm, but even so in 1920 was elected president of the American Gynecological Society.[47] Some American gynecological textbooks continued to oppose artificial insemination, but increasing numbers accepted it as a means of circumventing impediments to conception.[48]

Germans discussed the issue. On June 13, 1912, under the headline "Life Artificially Created," the *New York Times* reported that Albert Döderlein, a celebrated Munich professor, had announced the artificial impregnation of a twenty-four-year-old married woman.[49] The news precipitated in Germany a spate of publications musing about the impact of this technique on marriage and gender relations.[50] Following the war, the lively debates over almost every aspect of sexuality took place under the German Republic. Magnus Hirschfeld's Institute for Sexual Science sponsored an international conference from September 15 to 20, 1921, that included discussions of birth regulation through contraception, abortion, artificial insemination, and eugenic sterilization.[51]

So there is abundant evidence that artificial insemination of humans was discussed on the continent. What was being said about the subject in Britain? On first glance it appears that the answer is "nothing." Much more was said about the breeding of animals. Discussing primarily rabbits, dogs, and horses, the Cambridge scientist Walter Heape stated in passing that though human artificial insemination was a very simple undertaking, the chief obstacle was that the woman's ovulation cycle had yet to be accurately determined.[52] When the *British Medical Journal* reported in 1898 on the Vatican's condemnation of the practice, it primly observed that "artificial fecundation" was primarily discussed by French doctors. In Britain, it was not viewed as "within the sphere of practical medicine."[53] "This has been much practised abroad," stated a 1907 gynecological textbook. "In England it is not, or is very little, used, not only because it is distasteful, but because it is so generally a failure."[54] If one goes through *Index Medicus* and the *Quarterly Cumulative Index* for the first half of the twentieth century, one finds cited each year one or two journal articles on the subject from researchers in France, Germany, Poland, Czechoslovakia, the United States, Mexico, and Hungary. According to Schellen, it was not until 1943 that "old conservative Britain" contributed to the discussion.[55]

So where did Lillian Orford get her ideas from? The fact that mainline British medical journals did not devote attention to artificial insemination does not, of course, prove that it was not being discretely practiced and discussed. These same journals gave almost equally short shrift to the unseemly subject of birth control, and it certainly was being employed. And predictably enough, British sex radicals, who warned about the dangers of uncontrolled fertility, were the very ones who were most likely to inform the public of ways of combating infertility. Havelock Ellis, the best-known British sexologist, ended his discussion of sterility in *Studies in the Psychology of Sex* by remarking that "the attempt is occasionally made to effect artificial fecundation by injecting the semen from a healthy male." Ellis was clearly not opposed to artificial insemination, even by donor, but because he maintained the old idea that female orgasm was necessary for conception, he believed that these attempts would rarely succeed.[56]

Marie Stopes, the pioneering birth control advocate, was more optimistic. She enjoyed an enormous success with her first marriage manual, *Married Love: A New Contribution to the Solution of Sex Difficulties.* Appearing in 1918, it was an overnight sensation, going through innumerable printings and eventually selling more than a million copies. Her main argument was that the married woman had as much right to sexual pleasure as her husband.[57] But in correspondence, Stopes also showed sympathy for unmarried women deprived of the joys of motherhood, a state exacerbated by the loss of so many men in the Great War.[58] In the seventh edition of *Married Love,* Stopes went so far as to suggest recourse to artificial insemination by donor. She justified herself as follows:

> I have received an interesting series of correspondence from Australia, where under the name of "Scientific Motherhood," some high-minded women have been endeavouring for some time to found an institute for the scientific insemination of women war-deprived of mates, so that though husbandless they may have the joy and sacrifice of child-bearing under properly protected conditions. Although a hundred questions about this scheme arise unanswered to one's mind, it should be watched with special interest as it is planned from high motives so different from the unreasoning and selfish

> ones which have for too long added to the population in ways outside monogamic marriage to which we are callous because accustomed.[59]

Stopes's Australian corespondent was Marion Louisa Piddington. Piddington had attended the International Eugenics Congress in London in 1912 and by 1916 was arguing that in order to counter the dysgenic results of the war, widows and single women unable to find mates should be offered artificial insemination.[60] She won Stopes's support for the cause of "celibate motherhood."

In the 1918 American edition of *Married Love,* Stopes gave an extended defense of artificial insemination. She noted that customarily the woman was blamed for a barren marriage, but it was now recognized that the fault was often the man's, "particularly where the husband is a brain-worker in a city." A sterile man could prove his generosity by accepting his wife being inseminated by donor sperm. To spare his feelings, she suggested the injection be performed by a woman doctor. Stopes claimed that the success rate was good. Moreover she asserted that the old idea that "the soul and character of the child" was profoundly affected by the mother was now reinforced by what one knew of hormonal influences. And to win over husbands, Stopes argued that mental influences were also important. "Hence I suggest that the husband who is deprived of normal fatherhood may yet make the child of his wife's body partly his own, if his thoughts are with her intensely, supportingly and joyously throughout the whole time of the unborn baby's growth. If he reads to her, plays beautiful music or takes her to hear it, and gives her the very best of his thoughts and aspirations, mystical though the conclusion may seem, he does attain an actual measure of fatherhood."[61]

Medical men were not impressed by Stopes's reasoning. Her comments were subject to attack as unscientific and unsubstantiated, and the passages edited out of later English editions.[62] Nevertheless the incriminating lines were used against Stopes in the 1923 libel trial she launched against Halliday G. Sutherland, a Roman Catholic doctor who accused her of experimenting on the poor.[63] Sutherland's counsel cited her discussion of artificial insemination as evidence that her book would incite immorality. When Dr. Meredith Young, the medical officer for Cheshire, was testifying as to the morality of *Married Love*, he was asked by Sutherland's counsel why readers should be exposed to unsavory subjects:

> Do you think that artificial insemination should be introduced to them?
>
> It is a necessary corollary of the book.
>
> Is it not likely to arouse wrong thoughts in the minds of readers?
>
> It does not suggest that at all.

Do you as a doctor approve of artificial insemination?

It is for the woman herself.

Would you, as a doctor, practise it if a woman came and asked you if you would do it?

Yes, I would, with proper precautions.

Have you ever been asked to do it?

No; I doubt whether any medical man has been asked to do it in this country. It is better known in America.[64]

Although she ceased to write about the topic, in private correspondence Stopes suggested a friend or relative should act as donor, and that after the injection a wedge of cotton wool be inserted.[65] In America too, those who supported birth control were also likely to support treatments for infertility.[66]

Who were the British doctors who provided insemination services? It is difficult to say. We do know, however, that they included the Australian sex radical Norman Haire who practiced in London from 1919 on. He was active in the first British birth control clinics, a member of the British Society for the Study of Sex Psychology, the International Medical Group for the Investigation of Birth Control, the Eugenics Education Society, and helped organize the third congress of the World League for Sexual Reform that met in London in September 1929.[67]

Haire looked forward to a eugenics society, in which every medical school would have a chair of sexual science and every city an Institute of Sexual Science as in Berlin. For Haire, some should be prevented from breeding. "No crime is greater than that of bringing a child into the world handicapped from its birth by the inheritance of actual disease, deficient resistance to disease, or deficient life-energy."[68] But on the other hand in the future some would be selected as breeders and a certain amount of reproduction would be artificial. "This will probably be increasingly employed for impregnating women who do not desire to mate in the ordinary way, and for obtaining large numbers of children from especially desirable fathers."[69]

Haire later reported that Jervois Aarons and a number of other London doctors provided artificial insemination on "an extensive scale."[70] Dr. S. Jervois Aarons (d. 1923) of the Hospital for Women, Soho Square, and author of *Gynaecological Therapeutics* (1910) had a patient list that included London's elite.[71] Lady Diana Cooper later related that the term "Aarons' babies" was applied to those born as a result of his help. The joke, said Cooper, was that the babies were all his. "He may have had a stud of handsome footmen lined up outside, but he claimed that they were a genuine case of artificial insemination."[72]

References to women interesting themselves in such services even appeared in fiction. Virginia Woolf's short story "A Society" (1921) presents a group of young women voicing radical views:

> "Oh, bother," said Judith, who had been enquiring into scientific matters, "I'm not in love and I'm longing to explain my measures for dispensing with prostitutes and fertilizing virgins by Act of Parliament."
>
> She went on to tell us of an invention of hers to be erected at Tube stations and other public resorts, which, upon payment of a small fee, would safeguard the nation's health, accommodate its sons, and relieve its daughters. Then she had contrived a method of preserving in sealed tubes the germs of future Lord Chancellors "or poets or painters or musicians," she went on, "supposing, that's to say, that these breeds are not extinct, and that women still wish to bear children."
>
> "Of course we wish to bear children!" cried Castalia, impatiently. Jane rapped the table.
>
> "That is the very point we are met to consider," she said. "For five years we have been trying to find out whether we are justified in continuing the human race. Castalia has anticipated our decision. But it remains for the rest of us to make up our minds."[73]

In *Lady Chatterley's Lover* (1928), D. H. Lawrence made his distaste for artificial insemination clear by having it espoused by unsympathetic characters. Clifford is convinced that to maintain family traditions he needs a male heir. Connie is initially appalled to hear him say that he is willing for her to have a child by another man, but later toys with the idea. In the meantime, at one of their house parties the discussion turns to procreation. One woman is described as "reading a book about the future, when babies would be bred in bottles, and women would be 'immunized.'" She likes the idea as it would free the woman who would no longer be "dragged down by her *functions*." Clifford is attracted to a process that would overcome physical disabilities. "All the love-business, for example; it might just as well go. I suppose it would if we could breed babies in bottles." He likes the notion of repressing the needs of the body, whereas his guest hopes science will provide more freedom. "Only hurry up with the breeding bottle, and let us poor women off." Later in the novel, the talk shifts from ectogenesis to artificial insemination. Clifford muses that though impotent he might still father a child. He says to Connie, "Of course, *I* may have a child yet. I'm not really mutilated at all. The potency may easily come back even if the muscles of the hips and legs are paralyzed. And then the seed may be transferred." Mrs. Bolton, his caregiver, hearing of the conversation, thinks, "Yet doctors could do such things nowadays. They might sort of graft seed." Connie is once again outraged by such an unnatural idea. Indeed when she and Mellors are having their affair neither takes "precautions," though he suspects she's using him to produce an heir.[74] There's no doubting Lawrence's support for letting nature have its way.

In *Lysistrata; or, Woman's Future and Future Woman,* the misogynist Anthony Ludovici sketched out the nightmarish scenario of science being turned to the purposes of sex-hostile British feminists. "When once artificial impregnation is an every-day occurrence, a Parliament of women will doubtless pass legislation to make it illegal for any man to procreate a child naturally, if it is the wife's desire to have one by the intermediation of science."[75] In such a world, men would be reduced to the role of walking sperm banks. In *America the Menace: Scenes from the Life of the Future* (1930), the French writer Georges Duhamel sketched out a fantasy of American women selecting healthy sperm from a scientific institute with the same care as they chose the style and color of their car.[76] In December 1930, the Alhambre Picture Theatre in Leicester Square screened the American film *Just Imagine*. Described as "a merry musical romance of life fifty years hence," this comic film portrayed a future world in which women could obtain babies from vending machines.[77]

According to Lesley Hall, the sex radical Stella Browne went so far as to argue that lesbians had the "right to fulfill their maternal urges," but it is not clear if she was referring to recourse to artificial insemination.[78] We do know that some homosexuals thought of availing themselves of such services. In the early 1930s, the young Michael Tippett, the gay British composer, contemplated marriage and childbearing (by artificial insemination) with his bisexual friend Francesca Allinson.[79]

Stella Browne was the translator of Theodore van de Velde's *Fertility and Sterility in Marriage: Their Voluntary Promotion and Limitation* (1931), which contained a chapter devoted to both a history of and practical advice on artificial insemination. According to van de Velde, a Dutch gynecologist, women denied maternity fell into a "torturing state" that could have dangerous psychological and physiological ramifications. He, like Haire, believed that misfits (including homosexuals) should not reproduce, but he went on to indicate how healthy partners could use tampons, plugs of saturated cotton wool, or sprays of ejaculate to ensure conception.[80] By the 1930s, van de Velde claimed the success rate of this justified and legitimate operation was about one-third. Strikingly enough, van de Velde accepted the new notion that conception was most likely at midmonth but held true to the old idea that a woman's orgasm could assist conception. Accordingly a complex choreography might be followed. The couple first engaged in passionate lovemaking. The doctor then entered the room, syringed up the sperm from the "vaginal vault," and reinjected it into the uterus.[81] The process would be repeated each month for up to six months. Such a scenario was not always followed. Masturbation by the husband was offensive but might be required if penetration was not possible. Van de Velde also defended artificial insemination by donor, citing Hirschfeld to the effect that the process was altruistic, violated no ethical or eugenic standards, and was not comparable to adultery as it was impersonal "and not incompatible with marital fidelity and unity."[82]

From the mid-1930s onward, the British tabloids reported on "test-tube babies" in the United States.[83] By 1939, an account was given of a woman doctor in Manchester who, as a result of a simple "operation" with no anesthetic, was now the mother of a two-year-old. The paper stated that doctors were shy of reporting such cases.[84] A few marriage manuals referred to the process. Edward Griffith recommended artificial insemination using the husband's semen and a buffering solution.[85] A 1941 text advised its readers on how couples could practice artificial insemination themselves by inserting a semen soaked wad of cotton in the woman's vagina.[86] If artificial insemination was rarely discussed, it was partly due to British doctors regarding the practice as entailing crude and simple self-help practices that could not be dignified as scientific. On the other hand, though few defended the unseemly process of artificial insemination, many leading intellectual lights lauded the possibility of in vitro fertilization and ectogenesis.

Our first question was, how could Lillian Orford have known about artificial insemination when the practice was not being discussed in Britain? As we discovered, British medical journals might have shied away from the topic, but it was being broached by others. Like birth control, it was a subject in which patients were more interested than doctors and accordingly only mavericks within the profession defended it. The timing of the surfacing of such discussions is partly explained by the concern in the teens and twenties to make up for Britain's wartime losses.

Our second question is, what did the British medical profession make of this simple therapeutic strategy? It is hard to tell because between the wars British medical journals made only passing references to artificial insemination, usually drawn from American and European sources.[87] A full debate over the practice only took place in the 1940s when the use of donor sperm by the doctors Margaret Jackson and Mary Barton came to light. Suddenly the columns of the *Lancet* and the *British Medical Journal* were full of letters of protest.[88] Why was there such an uproar? One of its defenders, Kenneth Walker, noted that the public discussion of artificial insemination was "curiously remote" from its actual practice. Following up on his lead, one finds that the conversation about artificial insemination was not so much about the concerns for a basic medical procedure but for the ways in which it might have an impact on the medical profession, morality, marriage, paternity, and population.

The debate was begun in July 1943 in the House of Lords when Lord Brabazon took the occasion of discussing food supplies and the breeding of livestock to cite reports that in the United States thousands of couples had applied for artificial insemination.[89] If such services were made available in Britain, grave abuses would result, of which, in Brabazon's mind, one was particularly worrying: "Some women might prefer to have children without marriage, or by sinning." Other peers chimed in. Lord Glentanar lamented experiments in sex determination. Viscount Bledisloe—though he welcomed the importation of bull semen from the United States—fretted that tampering with human reproduction would result in the breakdown of family. The bishop of Chichester felt it necessary to declare that it was essential to draw clear distinctions between the breeding of humans and animals.[90]

As peers were renowned for parading their moral concerns, the discussion might have ended there, but Dr. Mary Barton, a physician doing pioneer work at the Fertility Clinic of the Royal Free Hospital, fired back a response in the *British Medical Journal*.[91] The medical profession, she tartly warned, could not take direction from the ignorant and the prejudiced on the courses of treatment to offer patients. Infertility posed serious problems.[92] Overcoming it resulted in happier marriages. Barton stressed that in most cases at her clinic the husband's sperm was used, but she let the cat out of the bag in admitting that in rare cases where the husband was sterile donor sperm was employed.[93]

Barton's letter precipitated a flurry of responses in the medical press. Doctors' first concern was that in carrying out a simple procedure they could become entangled in legal or even criminal proceedings. Such worries were exploited by opponents of artificial insemination like Dr. Letitia Fairfield, a Roman Catholic surgeon and sister of the novelist Rebecca West. Fairfield was well positioned to debate the matter, being qualified in both medicine and law, a suffragist, a Fabian socialist, and a veteran critic of eugenics.[94] She had written against birth control.[95] She had condemned sterilization.[96] Now Fairfield warned that artificial insemination posed grave legal implications. In theory, the law did not care how the conception of a married woman occurred, but Fairfield suggested that artificial insemination might not prevent a suit for nullity.[97] In short, one could have a situation in which a woman gave birth but whose marriage had not been legally consummated. What would be the status of the child? Fairfield again insisted that the use of "alien" or donated sperm raised problems. The resulting child was "clearly a bastard" though protected by the legal assumption that any child born to a married woman was legitimate.[98]

What was the status of the woman who employed donor sperm? The Orford case was cited by experts as showing that the courts regarded a woman's recourse to assistance without her husband's consent as constituting adultery.[99] But could both the doctor and donor be cited as corespondents?[100] And was the donor guilty of committing adultery against his own wife? Opponents also raised the specter of artificial insemination being used to commit fraud. Fairfield noted that insemination by donor could be used to defraud the heir presumptive of an estate. Or by a wife to dupe her husband. Or it could arm the donor with the means to blackmail the couple.[101] Even if consent forms were employed, courts might still view them as evidence of conspiracy to defraud. The issue was so sensitive that C. Gordon-Watson hoped that the General Medical Council would charge any doctor providing donor sperm with "unprofessional conduct."[102]

In response, the defenders of artificial insemination stressed that they took seriously their duty to judge a couples' worthiness, and they certainly did not service the unmarried, minors, or those who might use a birth to force a marriage or secure property. Barton admitted some women sought assistance to reassure a subfecund husband, or seek a forced marriage, or obtain a "guiltless conception," but they were not accepted for treatment. Legal precautions were taken. To protect doctors, the Medical Defense Union produced a consent form that was to be signed by all parties, including the donor's wife.[103] The Orford case was cited in a 1947 leading article of the *BMJ* that stated artificial insemination by husband posed no legal dangers but legislation was still needed to protect doctors using donor sperm.[104]

The issue, of course, only directly concerned a tiny number of practitioners. Though some attacked artificial insemination as a sign of scientific meddling, within the medical profession the elite regarded it as not scientific enough, as posing "no technical difficulty."[105] The process could work, but in the words of one doctor "because of the rather messy and almost sordid nature of the whole business, the method will not appeal to many women, and will rarely be urged by the physician."[106] Moreover male infertility problems, associated as they were with such unsavory subjects as masturbation and venereal disease that quacks customarily exploited, held out little attraction for ambitious specialists.[107] The fact that artificial insemination could be carried out by the low-status general practitioner or indeed even attempted by the determined patient meant that gynecologists regarded the operation as a demeaning exercise that they disdainfully left to others.[108] Being driven by patient demand, some saw it as threatening the authority of the profession. Just as specialists were largely indifferent to birth control until hormone contraceptive pills replaced floppy rubber goods, so too their interest in infertility only blossomed when the lowly syringe was elbowed aside by high-tech in vitro fertilization techniques.[109]

What light did the debate cast on the linking of morality and medicine? The most predictable opposition to artificial insemination came from Catholics like Fairfield.[110] The archbishop of Westminster appealed to Catholic doctors to oppose "test-tube" babies and reminded them of the church's 1897 condemnation.[111] Dr. H. P. Newsholme, a distinguished Catholic doctor, and one-time chief medical officer of health for the city of Birmingham, declared that homologous insemination violated aesthetics if not morality—and then only as long as no masturbation by the husband was involved—but no doctor could collude with a couple by offering them artificial insemination by donor.[112] To do so would be to degrade marriage.[113] Attacking both birth control and artificial insemination, one optimist enjoined his fellow doctors to teach the young to marry, make love, and "welcome gratefully the miraculous result!"[114]

Other doctors made vague appeals to morality. One—apparently forgetting about blood transfusions—declared it "wrong to introduce into the human body living organisms except in the natural way."[115] Another held that no decent couple would do it.[116] C. F. Keane condemned Barton's writings as "diabolical" and expressed his outrage that the *BMJ* allowed to appear in its columns a defense of "the pagan way of life."[117] A colleague saw polygamy as the next step.[118] Medicine, insisted another writer, necessarily involved morality, and doctors could not support "extramarital artificial insemination."[119]

Defenders of artificial insemination did not allow the moralists' attack to go unanswered. In response to appeals to the "natural," one correspondent pointed out that not so long ago anesthesia had also been condemned as "unnatural."[120] A Birmingham psychotherapist asked why Catholics like Newsholme condemned masturbation when it was employed by a husband to help his wife conceive. Was not such an act a sign of love rather than of perversion?[121] A woman correspondent more vigorously insisted that the issue should not be judged on "grounds of narrow-minded puritanism or bigoted Catholicism."[122] Given that since 1930 the Anglican Church sanctioned birth control in marriage, Kenneth Walker argued that there was no reason for it not to proceed to accept artificial insemination by donor, which similarly split sex and reproduction.[123]

Turning to the couple, both sides presented themselves as defenders of monogamous marriage. Opponents claimed artificial insemination led to the breakdown of marriages. Even the well-known sex expert Eustace Chesser, who carried out both AID and AIH, argued that the former practice could psychologically damage a family. The wife was prone to morbid musings, the husband lost self-esteem, and the child was most likely to be psychologically harmed.[124] A woman doctor added that she was convinced that the donor's relationship to his own wife was also bound to suffer.[125] Defenders countered with the argument that artificial insemination actually saved marriages. Margaret Jackson, who did pioneering work on using donor sperm at the Exeter Family Planning Clinic and Infertility Clinic, stated that AID was quite rare and usually only turned to when a marriage was breaking down.[126] In response to Chesser, she maintained that marriages were strengthened rather than weakened as a result of such infertility treatments.[127]

Both sides were pro-natalist and presented equally gloomy portraits of the infertile spinster. The needs of the unmarried had to be addressed by AID, argued Anne Ethel McCandless, so they would not end up as "embittered spinsters." The churches should accept that artificial insemination did not entail immorality, and the legitimacy of resulting children should be granted.[128] The duty of making up for wartime population losses was used to justify the most extraordinary suggestions. In the 1944 Medico-Legal Society discussion of artificial insemination, Judge Earengey airily voiced the opinion that some of the two million "surplus" women in Britain would likely take advantage of the method.[129] Dr. Marie Stopes reminded him that she had popularized the idea back in 1918, and that in Australia after the First World War this type of argument had been made but few spinsters had been interested.[130] Nevertheless in a 1945 issue of the *BMJ,* a eugenically inspired contributor continued to argue for AID for women without mates—spinsters and "young widows of good type" with as yet only one or two children. "It is greatly in the national and their own interest that these women should have further children, at least up to the level of four, which would be replacing themselves and their dead husband."[131]

The idea that unmarried women might have recourse to artificial insemination naturally appalled conservatives. Letitia Fairfield said it was outrageous to think the single would attempt it. No normal woman would bear a child whose father she did not know. Throughout history such women had been rightfully disdained by the community. "And who in that audience would like to think, as he grew up, that he was the child of a spinster and a syringe!"[132] Opponents implied that this was the goal of the infertility clinics; in fact, all the British doctors who provided artificial insemination were careful to underline the point that they never offered such services to the unmarried.[133] Prior to the 1950s, a few unmarried women who wanted to be artificially inseminated sought the help of C. P. Blacker, head of the Eugenics Society. He appeared to regard such requests as symptoms of mental instability.[134]

Both sides presented themselves as seeking to protect the family, but did they envisage any change in the roles of husband and wife? The opponents of artificial insemination warned that it could not help but sever the linkages of sex, marriage, and paternity. Its defenders insisted that it would not. First and foremost it served the interests of wives and would-be mothers. Joan Malleson argued that it was important to seriously appreciate the needs of the barren. For some women, adoption simply could not replace the experience of bearing a child. For such women, the insemination procedure was no more humiliating than any other gynecological process.[135]

Though they presented themselves as protecting women from intrusive medical interventions, a common charge leveled by opponents was that women who sought artificial insemination probably had neuropathic tendencies.[136] Others saw them as motivated by feminism.[137] It was thought not by chance that in the United States where women were freest artificial insemination was most available. In the 1920s, the misogynist Anthony Ludovici predicted that feminists would prefer "artificial impregnation" to natural intercourse. He returned to the topic in the midst of the *BMJ* debate, raising the fear that secret inseminations would lead to a subsequent generation ignorantly engaging in incestuous unions.[138]

Given the fact that women took the lead in the British discussions of artificial insemination, one can see why their opponents might have regarded it as part of a feminist plot. In the medical world, its most vocal proponents were Drs. Mary Barton and Margaret Jackson. It was defended by well-known birth controllers such as Marie Stopes and Joan Malleson.[139] In the medical profession, Beatrice Dobbie curtly asserted that in specific circumstances "the rational treatment is artificial insemination."[140] In Parliament when conservative MPs tossed about such terms as "disgusting" and "common decency," it was women like Dr. Edith Summerskill and Viscountess Astor who assumed the task of advising their male colleagues to try to be a little less emotional.[141]

How were men viewed in this debate? Opponents of artificial insemination felt no hesitation in portraying the donor as an obvious pervert.[142] One correspondent described the new specialists in "sipho-seminology" as almost as despicable as the degraded donor with his "perverted philo-progenitive cravings" that he kept hidden from his own wife.[143] Masturbation, still regarded by many as a perverse act, was obviously a fixation for such critics.[144] H. M. Stanley Turner was apoplectic that there were doctors who were not ashamed to defend the notion of a man masturbating for a fee and a couple raising the "adulterous offspring of a stranger!"[145] One writer claimed that patients were shocked by doctors seeking to recruit them as donors.[146] Another fantasized about the man "who hawks his seminal fluid round the countryside at so much per c.c."[147] Such lurid accounts, replied Margaret Jackson, were pure fantasies. She likened her donors, who were not paid but motivated by generosity, to those who gave blood.[148] Joan Malleson dryly lamented the fact that so many squeamish correspondents dogmatically appealed not to science but to religion or aesthetics, and clearly harbored "some vague ethical horror of semen being emitted anywhere other than in the vaginal canal."[149]

How did the role of husband emerge from such discussion? The need for artificial insemination obviously highlighted male inadequacies. It threatened to split social and biological fatherhood, and possibly blur the boundaries between marriage and adultery. For all these reasons, opponents declared the procedure to be morally wrong. Moreover gender roles were reversed with the husband being passive and the wife active. Such appeared to have been the situation in the Orford case. When Frederick proposed marriage, he was reported as suggesting that Lillian "make him intoxicated and then take him and marry him."[150] One of Lillian's letters was cited in which she wrote, "I wish I were a man. I would just be as keen and level headed as the next man."[151] Some twenty-five years later, opponents of artificial insemination continued to present themselves as seeking to protect the husband from being either victim or dupe. Letitia Fairfield cruelly sketched out a scenario in which a soldier might return home to find his pregnant wife protesting, "This was not an American soldier, it was artificial insemination."[152] In the war years in which Britain was haunted by the image of mutilated if not emasculated veterans, Fairfield's sally must have struck home.

The defenders of artificial insemination responded that, far from being displaced, husbands now had an even more important role to play in the family. Until the twentieth century, in cases of barren marriages doctors regarded men as innocent till proven guilty; women guilty till proven innocent. With new research, it was increasingly realized that husbands were responsible in perhaps a majority of cases.[153] "For centuries woman has borne the blame and burden of the barren marriage, and surely," argued Mary Barton, "it is time man accepted his part of the responsibility."[154] But defenders made a point of not simply challenging husbands but reassuring them. Great lengths were gone to in sparing the husband's feelings.[155] The secrecy in which the procedure was cloaked offered some protection of his manhood.[156] And the very fact that the procedure was called "artificial" insemination was used to underline the fact that it did not take the place of a man's love for a woman. Jackson stated that most requests for AID came from the husband.[157] Such an act was an additional proof of his courage and manliness. And thanks to AID, asserted Joan Malleson, the husband's self-confidence would grow and his marriage would be strengthened—not destroyed as opponents might claim.[158]

Finally, if the ways in which infertility was treated affected men and women, morals and medicine, they were also regarded by many as ultimately affecting the British nation as a whole. It says something of the pervasiveness of eugenic thinking in the 1940s that racial arguments were not only used to oppose artificial insemination; they were also used to defend the practice. Opponents argued that artificial insemination symbolized the undermining of civilization by materialistic science, and in particular by eugenics.[159] In the war years, such worries were very much in the air. The newspapers intermittently reported on the Nazis' experiments on breeding.[160] George Orwell exploited the theme in his classic dystopia *Nineteen Eighty-Four* (1949). He portrayed a totalitarian society where the hope was for "all children to be begotten by artificial insemination (*artsem,* it was called in Newspeak) and brought up in public institutions."[161] Orwell had been preceded by opponents of artificial insemination who presented it as a dehumanizing eugenic procedure. One critic chided Barton for forgetting that men were not cattle. On moral grounds one had to oppose "human stud-farming."[162] "One hears a great deal nowadays about 'State control,'" wrote a correspondent from Stoke on Trent. "Surely it is a degradation, and not a social advancement, to suggest that the population shall in future be organized into one vast stud farm?"[163] F. M. R. Walshe took the gloomy view that new therapies could result in "the making of a stud book, and the State registration of semen donors."[164] In reply, a "Service Doctor" wrote to protest that the reality was that only a tiny number of desperate couples sought fertility services. He cogently concluded, "Lurid pictures of vast stud farms and the imaginary scenes within are the product of a prurient rather than a reasonable mind."[165] But opponents of artificial insemination also used the opposite argument—that artificial insemination had to be prevented because its effects would be dysgenic. C. Rickword Lane protested that medicine already made possible the reproduction of mental defectives and diabetics; it should not add the sterile.[166] A colleague added that only "certain primitive types of women" would accept insemination, just the type one did not want to breed.[167]

When the *Lancet* reported in 1943 that artificial insemination was being debated in the House of Lords, it also noted that in the United States the National Research Foundation for the Eugenic Alleviation of Sterility was claiming that thousands of women had been helped.[168] The foundation's medical director, Frances Seymour, convinced of the racial improvements she had overseen in the United States, went on to float the idea of sending vials of American sperm to help repopulate war-ravaged Europe.[169] Even the Eugenics Society head, C. P. Blacker, described the scheme as absurd.[170] Nevertheless the incident was enlightening. Like their opponents, defenders of artificial insemination also wielded eugenic arguments, presenting themselves as dealing, not simply with private dilemmas, but with the nation's population problems. "Actual individual sterility and subfecundity, both male and female, are on the increase," argued Barton in justifying her studies. "The urgency of the problem of combating the falling birth rate should always be present in our thoughts when weighing this question from any angle."[171] Speaking before the Medico-Legal Society, Robert Forbes concurred that "any process that proffers assistance" to population growth deserved a hearing. Some demographers were actually predicting depopulation. In discussion, Judge Earengey agreed that artificial insemination could play a role in countering the "falling birth rate," and fill "some of the empty places in the Empire."[172] C. O. Carter argued that Britain needed the rapid production of children with good genes. Above-average couples should have six to eight children, the average four.[173]

From the 1920s onward, a number of writers had defended artificial insemination as a form of positive eugenics.[174] In 1935, Herbert Brewer, socialist and postal clerk, argued that race improvement could be achieved by controlled breeding, all couples selecting sperm that had been determined to have the best possible hereditary endowment. Although men would no doubt initially oppose being displaced by sperm banks, Brewer believed they would eventually be won over.[175] The biologist Cyril Bibby agreed that it made eugenic sense for more than one woman to avail herself of superior sperm but warned that there were dangers. "The prospect of thousands of little Adolphs is appalling."[176] Though the geneticist Herman Muller did not believe that Brewer's notion would work under capitalism, under socialism he envisaged donor insemination being practiced. "Only social inertia and popular ignorance now hold us back from putting into effect (at least in a limited experimental way) such a severance of the function of reproduction from the personal love-life of the individual."[177] He assumed that an enlightened woman would be proud to bear the child of a Darwin or Lenin. Julian Huxley agreed that with the availability of birth control and artificial insemination it was now possible to separate the sexual and reproductive functions. "The perfection of birth-control technique has made the separation more effective; and the still more recent technique of artificial insemination has opened up new horizons by making it possible to provide different objects for the two functions. It is now open to man and woman to consummate the sexual function with those they love, but to fulfill the reproductive function with those whom on perhaps quite other grounds they admire."[178] In each scenario, the talk was of passing on the traits of the "superior" male; women's endowment was judged to be not that significant.

Medical defenders claimed that AID could be used not just when the husband was sterile but when his genetic disposition raised the risk of such complaints as deafness or nervous disease.[179] Barton made a point of stressing that she sought superior donors. They should be mature and already fathers, but the doctor would consider as well the "eugenic quality of the donor's stock."[180] Dr. Margaret Jackson, of the Exeter Family Planning Clinic and Infertility Clinic, was both a gynecologist and member of the Eugenics Society. She argued that AID offered a better selection of genes than a fling with a "fancy man." Fertility panels were choosy about donors, and in the future she predicted the Medical Research Council and the Eugenics Society could give advice on their selection.[181]

Some supporters of Barton employed the crudest eugenic rhetoric. One doctor asserted that artificial insemination was preferable to adoption because of "the frequently poor genetic quality of the material presented for adoption." Another protested that it was necessary to counter the current threat of "racial degeneration" in which the country tolerated the breeding of "any and every sort of rubbish" by turning to means of race improvement.[182] Recalling the eugenic language in which artificial insemination was described and rationalized does not diminish the courage that pioneering doctors like Mary Barton and Margaret Jackson needed to help women desperate to become mothers and advance an initially unpopular cause. What it does do is remind us of how difficult it was in the first half the twentieth century to speak of the importance of reproduction without falling back on old platitudes about the need to "improve the race."

In the 1940s, nothing could rival artificial insemination in better representing reproduction by design. In so doing, it dramatically demonstrated the splitting of sex and reproduction. For this reason, the reassurances offered by its defenders are all the more interesting. Downplaying as much as possible its potential impact on attitudes toward gender and paternity, they insisted that the procedure only assisted men and women become what nature intended them to be—mothers and fathers. Far from overturning traditional gender roles, a simple service would allow women to become maternal, men to be manlier, and families to be fruitful and fulfilled. Drawing on eugenic notions, the argument was made that British couples, the British nation, and the British race would all benefit. Of course, in the long term artificial insemination would also be employed by single women and lesbians to radically redraft our notions of the family.[183] In the short term, that is to say in the first half of the twentieth century when the procedure was only emerging, it was predictable that that its defenders would seek to portray it in the most conservative of hues.

PART THREE

Romantic Racialism

CHAPTER 6

"Breeding a race apart from nature"

Ruralists and Conservationists

In Huxley's *Brave New World,* citizens are conditioned to abhor nature. The director of hatcheries and conditioning explains: "A love of nature keeps no factories busy. It was decided to abolish the love of nature, at any rate among the lower classes. . . . But simultaneously we condition them to love all country sports. At the same time, we see to it that all country sports shall entail the use of elaborate apparatus. So that they consume manufactured articles as well as transport."[1] Like many interwar writers of scientific romances, Huxley used his bleak portrayal of a future overurbanized world to implicitly highlight the values of contemporary country life threatened by the forces of modernization. An anonymous reviewer of the *New Statesman* described the novel as boring and repetitious yet conceded that it would please the defenders of the "sacred peasant." "Any life, however dung-scented, fly-blown and fever-ridden, would seem sweet compared with the drugged, unimaginative, mechanical and conditioned boredom of the perfectly happy inheritors of the kingdom of our Ford."[2]

Huxley reminds us that in the first half of the twentieth century those who discussed the merits of planned reproduction were not simply thinking of the multiplication of machines and humans. The question was posed if the reproduction of nature itself—in particular its forests and wildlife—was assured. Of course lamentations for the ending of country life had a long history in Britain.[3] What was new were the emerging conservation and environmentalist movements that declared that the threats posed to the natural world were now so great that it could not be left to fend for itself; rational direction had to be given to its protection. Moreover they argued that the breeding of healthy humans and healthy plants was inextricably linked, an assertion that clearly had a political resonance.[4] They produced their own futurist literature, warning that natural catastrophes would ensue if the right sorts of measures were not taken in time.

A number of groups voiced such holistic concerns. The Men of the Trees was in many ways representative of such organizations. I first came across references to the Men of the Trees in the private papers of one of the more extreme eugenicists of the interwar period, Reginald Ruggles Gates. A Canadian-born botanist, Gates worked mainly in England where he established himself as an expert in cytology and genetics. Despite his eminence, he was one of the few twentieth-century scientists who maintained the idea that there had originally been more than one human species and that race crossings had harmful effects.[5] His papers also include some letters and circulars from the Men of the Trees. The name conjured up Tarzan-type figures, but it was in fact a British forestry society dedicated to the planting and protection of trees. At first glance, it might seem surprising that Gates was a supporter and its treasurer was Ursula Grant Duff.[6] Duff was the English Eugenics Society's chief "society" organizer. Like Gates, she too was much preoccupied by the purported danger of "race suicide." She supported sterilization of the feebleminded and in 1935 attended the Nazi exhibit in Düsseldorf on "race knowledge."[7]

Why were eugenicists like Duff and Gates involved in the apparently benign Men of the Trees movement? The short answer is that both eugenicists and environmentalists embraced the notion of reproduction by design. The long answer first requires an examination of the life and works of the movement's founder, Richard St. Barbe Baker, and then a survey of the activists who shared his views. Such an investigation reveals why the histories of eugenics and environmentalism were curiously entangled. In what follows, the reasons will be teased out why those in the interwar years preoccupied with protecting racial stocks and those seeking to protect native species were drawn together.

Conservationist, vegetarian, supporter of the Baha'i faith—at first glance, Richard St. Barbe Baker appears to have been a sort of secular saint.[8] His main claim to fame was being the founder of the Men of the Trees association. It still exists today as the International Tree Foundation. Born in England in 1889, he proudly traced his family's stock back to the time of Henry I. His interest in trees came early as his father, a minister, had a nursery—indeed he had two varieties of apples named after him.[9] Baker was also marked by his father's religiosity and admiration for colonials, or as he put it, "men of our stock who had blazed the trail of overseas settlement."[10] Accordingly Baker set off for Saskatchewan in 1909 to begin divinity studies at Emmanuel College, affiliated with the University of Saskatchewan. But he also did some homesteading south of Saskatoon. He later asserted that it was witnessing farmers' thoughtless clearing of the dwarf willows and the resulting soil erosion on the prairies around Prince Albert that convinced him to go into forestry. He returned to England just in time for the outbreak of World War One and after having been invalided out of the army in 1918, completed his studies in forestry and silviculture at Cambridge.[11]

Before the war, the Germans and French led in forestry studies.[12] It was a relatively new field, but the British had begun a forestry service in India and the empire obviously offered the greatest scope for the innovative young forester.[13] In November 1920, Baker set out for Kenya where he was appointed assistant conservator of forests.[14] As a forest officer in equatorial Africa, he again witnessed the ravages of deforestation—due to the invasions of nomadic farmers, cruel grasses, and encroaching deserts.[15] In response, in 1922 he laid the foundations for the Men of the Trees association. Recruiting several thousand voluntary tree planters, he sought to instill in Kenyans the importance of reforestation. He supported similar policies when made conservator of forests in Nigeria in 1924, but following disagreements with superiors, he left the colonial service shortly thereafter. Nevertheless he continued to promote forestry and conservation, attending the first World Forestry Congress held in Rome in 1926.[16]

Baker might have remained little more than an obscure, itinerant lecturer, but luckily for him a visit to America rescued him from oblivion. To many in Britain, the United States represented all that was wrong with modernity. Philip Gibbs presented it as a self-centered, intolerant, and ruthlessly antilabor society.[17] For Joad, its most glaring failing was its crass materialism: "Hot baths and more hot baths, larger and ever larger hotels, faster and faster cars, golf played by ever fatter and more vulgar men, and lap-dogs kept by even fatter and more vulgar women, cocktails and culture, psycho-analysis and faith healing."[18] Aldous Huxley warned that America was the future, a land in which even young women appeared to be mass produced. On his first visit to Los Angeles, he was stunned. "Thousands and thousands of flappers, and almost all incredibly pretty. Plumply ravishing, they give, as T. S. Eliot has phrased it, a 'promise of pneumatic bliss.' Of pneumatic bliss, but of not much else, to judge by their faces. So curiously uniform, unindividual, and blank. Hardly more expressive—to the foreign eye, at any rate—than any of the other parts of that well-contoured anatomy which they are at such pains to display."[19] Fascinated by American urban life, such observers failed to note the importance of the nation's vibrant environmental and conservation movements. Baker did not.

Having been in Palestine in 1929, Baker arrived in the United States with slides and films to lecture on his tree-planting exploits.[20] In New York, he was discovered by the famous journalist and broadcaster Lowell Thomas, who a decade earlier had "made" T. E. Lawrence via his films *With Allenby in Palestine* and *Lawrence in Arabia*. In the interwar years, Americans had an insatiable appetite for such romantic figures, and Thomas now presented Baker as yet another remarkable man with a mission—dubbing him his "Lawrence of Africa."

The same year that he was discovered by Lowell Thomas, Baker met the leading American environmentalists. The best known were eugenicists.[21] He was introduced to Madison Grant, the author of one of the classic works of scientific racism, *Passing of the Great Race* (1916), a well-known eugenicist and conservationist. On the one hand Grant worked for immigration restriction and antimiscegenation policies, while on the other he pioneered innovations in conservation wildlife management.[22] Baker was contacted by Newton B. Drury of Save the Redwoods League. Its cofounders in 1918 were all eugenicist who advanced a racialized notion of the wilderness. Baker joined in their campaign to save the sequoia and visited California to see the majestic trees in person.[23] In a country where Anglo-Saxons were beginning to feel themselves engulfed by immigrants, the middle-class public proved responsive to men asserting the need to protect a "splendid race" of trees.[24] They won the financial support of philanthropists like the Rockefellers, and the redwoods were finally protected. Baker perhaps exaggerated the role he played in the campaign, but he did contribute to popularizing in England the iconic image of the endangered forest giants.[25]

While on the West Coast, Baker also made a point of visiting the garden of the famed horticulturist Luther Burbank, "father" of the Russet potato, optimistic Lamarckian, and eugenicist. Burbank, an expert in producing hybrid vegetables, also pontificated on raising humans. He declared "human seedlings" needed good food and exercise more than schooling, and produced an essay entitled "The Training of the Human Plant."[26] Baker was also impressed by the work of James G. Eddy on tree hybridization.[27] Eddy, founder of the Institute of Forestry Genetics at Placerville, California, would later shift his interest to the "human tree." In 1938, he sought funds from the Carnegie Foundation for a Clinic of Human Heredity.[28] Drawing on metaphors of pure blood and pure bodies, all these men argued in their various ways for the necessity of safeguarding America human and plant stock from foreign pollution.

Prodded by his friends, Baker cobbled together his lectures into a book, *Men of the Trees*. Thomas not only helped publicize the book on his NBC radio program, he also wrote its introduction.[29] There he described Baker (whom Thomas preferred to refer to as "Captain" Baker) as "the blue-eyed, ruddy faced, red mustached, British soldier type" who had an extraordinary passion for trees.[30] Thanks to such publicity, Baker's reputation was made. He won over wealthy and influential patrons in America and England, and the rest of his life was taken up by writing and lecture tours. The Men of the Trees association became an international organization and the main focus of his energies.[31]

Baker recognized the importance of fashioning an appropriate image of the gentleman adventurer. He had friends in the press who reported his exploits.[32] To popularize his message, he produced over two dozen books and became an active public speaker and broadcaster.[33] Most of his books simply retold in one form or another the story of his life. How reliable the accounts are is difficult to judge. He attributed to himself an importance in establishing the British Ministry of Health and the United States' Civilian Conservation Corps. Every reforestation program in every country he visited, he claimed, was due in part to his activities. A United Nations official later recalled that by the 1960s Food and Agriculture Organization officials regarded Baker as an eccentric and yet recognized that he had been one of the first to publicize the importance of reforestation and watershed management and the challenges posed by loss of biodiversity and climate change.[34]

Why in the interwar years were Baker's lectures and writings so popular in Britain and North America? A Royal Forestry Society had existed from 1882. Other experts in forestry had already produced academic texts on the need for conservation. In a work such as Augustine Henry's *Forests, Woods and Trees in Relation to Hygiene* (1919), for example, there was extensive discussion of tree shelters, afforestation of pit mounds, protection of water catchment areas, parks, and forests. Such scholarly works assumed that their readers would include the statesman, economist, engineer, doctor, and forester.[35] In contrast, Baker targeted a more popular audience. He was not indifferent to such issues as forest sustainability and efficiency, but he primarily invited his audience to participate in a moral crusade.[36] For example, he cited Ruskin to the effect that trees shaped the character of men.

Baker's romanticization was most clearly manifested in his anthropomorphic portrayals of the tree, which he called man's best friend and God's noblest creation. Only the cruel and ungrateful refused to recognize this relationship. It was natural to be a "tree lover," he wrote, since trees were more civil, "trusty friends" than many humans.[37] Accordingly the Men of the Trees association produced lavishly illustrated calendar books of trees in which their spiritual and aesthetic, rather than their economic value, was stressed. Pictures of trees were presented with such stirring titles as "An Ancient Sire" or the "Madonna of the Woods," and there followed rhapsodic accounts of their individual characters and personalities. The elm, we are told, was "friendly," the yew "reserved" (especially when "tortured" by topiary), the ash (the "Venus of the Woods") was beautiful, cedars liked companionship, and the English elm was declared more "solid" than its American cousin. Baker came close to sexualizing some species. He admitted to finding especially seductive the Douglas firs of Vancouver Island because in growing so close together they lost their lower branches, or what Baker called their "crinolines," allowing one to admire the trees' "naked stems."[38]

But back to our central question. Why would eugenicists like Gates and Duff be supporters of Men of the Trees? The mouthpiece of the English eugenics movement, the *Eugenics Review,* focused on such topics as marriage, reproduction, differential birth rates, poverty, crime, degeneration, insanity, and sterilization. It is true that its authors bandied about horticultural terms such as "stock," "breeding," and "hybridization."[39] They often used botanical metaphors when discussing human reproduction and cited works with evocative titles like *Figs from Thistles* and *The Fruit of the Family Tree.*[40] Nevertheless the contributors to the *Eugenics Review,* when talking about the environment almost always meant it in the restricted sense of the human or built environment.[41] They certainly did not discuss forestry. And at first glance it is difficult to detect anything in the Men of the Trees' calendar that would particularly interest the eugenically minded. But in closely examining what Baker had to say about empire and race, urbanism and class, one can begin to see why those preoccupied with protecting racial stocks would be drawn to a movement seeking to protect forests.

First, there was Baker's assumption—which he shared with most eugenicists—that Western, white experts knew best how to protect the world's natural resources—be they trees or people. Baker began his career in Kenya. By his account, the country was experiencing an ecological disaster, deforestation being caused by Arabs, whites, and in particular the natives.[42] He made clear his unthinking elitism by the patronizing language he employed. The "childlike, simple and impetuous" Kenyans, he reported, were destroying their forests, but acting out of sheer ignorance could not, of course, be punished.[43]

How could they be taught to plant trees? Baker first thought of inducing them to copy the Boy Scouts in doing good deeds, but accepted a brother colonial officer's advice that the idea of putting natives on their honor was "wildly impracticable and quite impossible."[44] Then Baker struck on the idea of organizing a tree dance to attract the attention of the locals. Promising an ox for the best male dancer, and a necklace for the best female dancer, he claimed that thousands showed up. He then took the opportunity to hector them on the dangers of deforestation. He called for volunteers to become members of an elite society called Men of the Trees, who vowed to protect trees, and were rewarded during an impressive initiation with a badge of office. "The whole ceremony has a tremendous effect upon the simple and impetuous heart of the African warrior," Baker recalled. "The idea of performing one unselfish act every day in the service of others was entirely new to the thoughtless pleasure loving warrior, and he did not quickly understand the idea underlying this pledge."[45] The ritual swearing, Baker claimed, revolutionized young men who had never thought before of being unselfish.[46] Thanks to white men, the African was trained to leave his "old improvident ways" and become "industrious in spite of himself."[47]

Baker's self-congratulatory narrative in effect contrasted "the dynamism of the colonial administrator with native fatalism."[48] His underlying argument was that nomadic farming was destructive of a natural system and that the introduction of new silviculture could strike a healthy balance. The importation of such new knowledge and expertise to the colonies was important not only in protecting the environment, but in impressing on the natives the abilities of white men. Baker in effect presented himself in his writings as yet another great white chief protecting the natives. Such displays of competence were vital, he stressed, because the Great War had badly damaged European prestige in the nonwhite world.[49]

Eugenicists took it as given that the medical scientist, the expert, had a duty to determine who should or should not reproduce. Similarly, for the colonial official, the concept of stewardship—of natural resources or subject peoples—had enormous appeal. Baker assumed that the history of the world was the history of the white man. When he provided a timeline of Kenya's history, he began it with the arrival of English explorers in the 1850s.[50] Similarly he stated flatly that the "solitude" of the California redwoods had only been broken when they were discovered by white prospectors.[51] Baker was willing to acknowledge that the arrival of whites in "virgin territories" often caused problems. Britain, which "opened up" foreign lands by clearing them of trees, had unthinkingly created potential environmental disasters. The British therefore now had the duty to instruct the natives in the need to reforest. But white men were still to direct the natives. So Baker relished his success in Kenya in having the Kikuyu follow his injunctions. He similarly praised General Allenby's tree-planting efforts in Palestine. On his way to Jerusalem, Baker happily reported, the general had seen to it that "any Arab loafers found in the villages were to be put to work, planting trees."[52]

Baker's hope to maintain the empire was reflected as well in his sharing the fear of other whites in Kenya that they would be swamped by Asians.[53] Though the latter far outnumbered English settlers, they initially had no elected representatives in the Kenyan legislative council. In Britain, Labour MPs were impressed by the argument that as the Indians had British passports they should have political representation, but Baker argued that the African population had to be "protected" from them and he claimed to have played a key role in the Devonshire Report of 1923, which postponed democratic reforms. He congratulated himself that as a result of his intervention he and his fellow British colonial officials remained the trustees of Africans. "We must," he argued, "fulfill our trust and not betray them (the Africans) to the Indians or the Europeans."[54]

At home, English eugenicists were haunted by the specter of their country being swamped by the riffraff of Europe. They were accordingly responsive to Baker's argument that British *native* plant species had to be protected from *alien* invaders. His friend Massingham condemned the fir tree as a Teutonic impostor and attacked the Forestry Commission (established in 1919) for imposing a coniferous monoculture.[55] Such apparently innocent assertions echoed the eugenicists' concerns about restricting immigration. Baker lamented the appalling desire of the English to plant exotics and enjoined his readers: "Let us try to keep the character of the country by planting the trees which belong to the country."[56] Native trees, he insisted, provided inspiration and spiritual support. "Trees are healthful, exhilarating, especially our native wild ones growing where the great creative force of nature placed them."[57] Yet at the same time, he contradicted himself in campaigning to have English oaks grown throughout the empire.

In making his nativist claims, Baker was conjuring up the romantic notion that an uncontaminated environment had once existed. Terms such as "native" and "alien" when applied to trees suggested that such categories were a natural given. The implication was that the land was once pristine, but in reality that had obviously never been the case. Constant changes and migrations of species were always taking place. In the rhetorical laments for the threat to native species posed by invasive newcomers, one can detect an obvious displacement at some level of current social preoccupations. Protection of habitat meant keeping aliens out.[58] It is difficult not to conclude that vague racist attitudes underlay such thinking.[59] Baker was certainly not alone in making such statements.[60] In Weimar and Nazi Germany, similar sentiments were expressed by garden writers who stressed the "rootedness" of native plants and the superiority of the wild German garden over the un-Nordic formal or French garden.[61]

Baker's views on class were also similar to those of many eugenicists. Protecting trees, he believed, had social and political consequences.[62] Forests protected the spirit of men and could save a nation. By this, Baker meant that thriving forests ensured a healthy rural population. If farmers left the land and headed to the cities, unrest inevitably ensued. Consequently he expressed the hopes that the radio would allow Africans to be educated without requiring them to come to cities. "Book-savvy" types, he asserted, never returned to agriculture.[63] He later was to claim that the Mau Mau failed to succeed in those very areas of Kenya where tree planting had taken place.[64]

And if Baker wanted to keep Africans out of the cities and on the land, he strove for the same goal for the English. Similarly, in much of eugenic popular writing, there was a distinct antiurbanism. The commonplace assumption was that racial decay was found in cities where men and women were less healthy than in countryside. In 1938, the journal of the Men of the Trees reported the death of Grey Owl, the Canadian First Nations conservationist and a member of Men of the Trees. In Ojibwa costume, he had lectured in Britain in 1935 and 1937, publicizing his book, *The Tree*. The journal took the opportunity of his death both to express amazement at his command of English and to contrast his wonderful physique—due to "the grandeur and the uplifting powers of the ennobling forests"—to the poor state of many who heard him lecture.[65] Unfortunately it was soon revealed that Grey Owl was in fact Archibald Belaney, a farmer's son from Hastings who, after immigrating to Canada and working as a trapper, successfully passed himself off as a native.[66]

Nevertheless Baker maintained his working premise that the city was bad and the countryside good.[67] He cited the Prince of Wales to the effect that "Fine forests make fine men."[68] At a Men of the Trees meeting, Archdeacon Lonsdale Ragg spoke of "the strengthening and beautifying of character that came from working among trees."[69] Man's primeval instincts would inevitably force him to return to the forests, insisted Baker; city life was fundamentally unbearable.[70] The large metropolis created by the Industrial Revolution was a mistake. "Our industrial age is admittedly not conducive to human dignity," he lamented.[71] It had produced a spiritless proletariat. He went so far as to assert that the English urban working class was worse off than the Kikuyu. "A return to slavery would have less horrors for the African native than the treadmill routine of factory wage-slaves who, having sunk their individuality in a number or a letter, ring on and off as if they were merely cogs in the wheel of a great heartless machine."[72]

When the Depression hit, hard-line eugenicists opposed welfare measures. Baker agreed that it was disgusting that "unemployed wage slaves" clamored for unemployment pay rather than risk going overseas to farm.[73] His answer to the social dislocations of the 1930s was to get men back onto the land. He told the story of one Welsh boy who came from a farming family but whose grandfather "had been tempted by big money to the mines."[74] Fortunately the boy was sent out to Canada, and, according to Baker, his fairy-tale success was proof positive that farming was in his blood, as it was in the blood of most healthy Britons.

During the 1930s, Baker campaigned to have Britain establish in the countryside Civilian Conservation Corps camps like those in the United States to teach the unemployed self-discipline and military training.[75] By such camps, he declared, Franklin Delano Roosevelt had saved "an army of idle youth from moral erosion caused by unemployment, while, on the other, he attempted to save his land from deterioration and barrenness."[76] In England, a generation of young people did not know their own native parks and woodlands. "How," Baker asked, "can native patriotism be expected under such conditions? What has the State done for them but kill their initiative by keeping them in idleness?" Forestry camps and conservation work would not only get youths off breadlines. Life in the countryside would teach the young "lessons not taught in towns."[77]

In the early twentieth century, many shared the notion that youth had to be rescued from the debilitating atmosphere of city life. The Boy Scouts movement was only the first and best-known organization to laud the rejuvenating influences of country life. It was followed in England by competitors such as the Kibbo Kift Kin and the Woodcraft Folk. All inculcated a reverence for the natural world and espoused vaguely eugenic notions that outdoor activities would lead to the breeding of a better race.[78] According to the founder of the Kibbo Kift Kin, John Hargraves, "Lamp-posts and pavements, motor-buses and tube stations are hardly likely to promote the sacredness of the realization of love and life which belongs instinctively to the woods, and scents, sounds, and sights of Nature."[79] To restore spiritual values, it was necessary "to get Earth contact in a mechanical age." Giggly girls, nagging women, and those with "sentimental attachments towards their own sex" provided an "exhibition of devitalization." What was needed was the "strong, potent, well-knit man," not the "gentle hairless men" and "simpering nymphomaniac women who would swallow up Priapus and still remain unsated."[80]

Eugenicists frequently complained that elected politicians were afraid to make tough decisions to protect and improve the population. Baker agreed. Looking back on the Depression years, he asserted that Britain wallowed in economic misery due to its lack of leadership, whereas Palestine prospered under the no-nonsense rule of British military officers.[81] He likewise repeatedly harped on the benevolent powers he wielded as a colonial official. In a book published during

World War Two, he even noted the flattering comparison made between his activities and those of the Fascists. According to Baker, when the Italian ambassador to Kenya saw the 1923 march of the Men of the Trees, he exclaimed, "Baker, this is marvelous! This is the Fascita of Africa. You must meet my friend, Mussolini." In 1926, they did meet in Genoa and talked forestry. The Englishman was impressed by the Fascists' realization that Italy badly needed reforestation. "These young men, I felt, were desperately in earnest and, and anxious above all to act, and act quickly, for the good of their country."[82] Mussolini later "recalled with relish that he had ordered the reforestation of the Apennines in order to render the Italian climate 'more rigorous' and produce a 'more thorough selection and the improvement of the race.'"[83] In the main, Baker presented himself as above politics. He preferred the forest that he described as the perfect self-regulating system, combining as it did the best of capitalism, communism, and Social Credit, but had no clerks or councilors, no taxes or debt.[84]

Who supported the Men of the Trees? Baker sought the support of the great and the good. They included a founding member of the Eugenics Education Society, C. W. Saleeby; the anthropologist George Henry Pitt-Rivers (grandson of the archeologist Augustus Henry Lane Fox-Pitt-Rivers); the Quaker welfare worker and philanthropist Elizabeth Cadbury; the Earl of Bessborough (Frederick Edward Neuflize Ponsonby); the agricultural reformer Sir Horace Plunkett; Lord Mamhead (Robert Newman, one-time Conservative MP for Exeter); Field Marshall Allenby; the suffragist and preacher Maude Royden; the Social Credit cultural critic and commentator Philip Mairet; the promoter of organic farming and a founder of the Soil Association Lady Evelyn Balfour; the architect of the British Union of Fascists' rural policy Jorian Jenks; and the Irish nationalist Shane Leslie.

A few individual supporters warrant special attention because they demonstrate the various ways in which concerns for the environment and fears of mass society were intertwined. In 1934, explorer, army officer, and theosophist Sir Francis Younghusband agreed to become first president of the British chapter of the Men of the Trees, a position he held until 1942.[85] He had been introduced to Baker by Ursula Duff and was an important catch. In 1903–4, he had led a British expedition into Tibet and in 1921 launched the first reconnaissance of Mount Everest, according to his self-serving account having had to push aside the fearful Sherpas.[86] The hero of *Lost Horizon* appears to be in part a fictionalized version of Younghusband. Due to his fame, in 1925 he was elected president of the Sociological Society, an organization that at the time was seeking to wed sociology, geography, mythic religions, and eugenics. In his presidential address, he made clear that his experience with primitive peoples convinced him that neither "level equality" nor any "free" individuals ever existed; there were always social gradations. What civil society needed was not socialism but the sort of interdependence found in a regiment or a mountain-climbing team. Not surprisingly he was initially attracted to Italian fascism and lauded Mussolini. "The world will go forward. Progress will be made. Perfection will be achieved. But not by all—not by sluggards. Sluggards—whether nations or individual men and women—will be dropped out, just as slugs have been left behind."[87]

Moreover Younghusband insisted that the most astute recognized that humans were bound together by both worldly interests and the force of a "Universal Spirit."[88] Drawn more and more to mysticism, in July 1936 Younghusband convened in London a World Congress of Faiths attended by Baker. They shared a spiritual view of nature expressed in Baker's words: "We shall enter the sanctuary of the woods as if treading on holy ground, seeing that we are in company with tree-beings who respond to our attitudes towards them."[89] At the same time, Younghusband harbored some quaint notions about the linkage of sex and nature. Though condemning homosexuality as a "cancer," he liked to climb in the nude.[90] Heterosexual love he described in mystical terms. It had to be both a spiritual and physical act. The British were, he wrote, the most poetic of races. Drawn together by their natural beauty, manhood and womanhood necessarily answered each other's call.[91]

A 1920 talk by Younghusband to the Royal Geographic Society on the need for geographers to study beauty inspired the geographer Vaughan Cornish to produce a series of works—similar to Baker's—aimed at educating the public on how to appreciate scenery.[92] Industrial towns he condemned as unsightly, whereas in nature one found perfect harmony. Indeed communing with nature was a form of worship. In mystical, rhapsodic language, Cornish described how nature brought peace and satisfaction, absorbing the individual in a timeless "cosmic union."[93]

Cornish linked racial identity to landscape, arguing that the English had a special perception of the beauty of trees.[94] Turning to Britain's overseas possessions, he produced the sort of imperial geography that wrestled with the question of whether whites could acclimatize to the tropics and settle "empty" lands.[95] He was happy to couch such discussions in eugenic terms. In a paper presented to the Eugenics Education Society in 1924, he asserted that geographers had no doubt that the white race added the most to the world's stock of knowledge. The problem was that it was outnumbered two to one. Race mixing was, of course, to be avoided as it resulted in a "chaotic personality." If the desired racial homogeneity in the dominions was to be assured, each couple would have to produce at least four children. Youth had to accept the responsibility of serving as the trustee of the race and nation.[96]

The novelist Vita Sackville-West sat on the Council of Men of the Trees from 1938 onward.[97] Best known today as the lover of Virginia Woolf and the wife of Harold Nicolson, she prided herself on her writing. In 1946, she was made a Companion of Honour for her services to literature. The following year she began a weekly column in the *Observer* called "In Your Garden" that popularized the Sissinghurst style (named after her estate). In 1948, she became a founding member of the National Trust's garden committee and owed her public reputation to her gardening expertise. Less well known was the fact that she was an avowed eugenicist. Predictably enough she hated the Beveridge Report because of its welfare provisions, writing to Harold, "I think it sounds dreadful. The proletariat being encouraged to breed like rabbits because each new little rabbit means 8/- a week."[98] Her fiction was rabidly anti–working class. In *Heritage* (1919), her first novel, she cited Galton in telling the tale of woman whose caddish husband—"he slunk about like a Jew"—eventually goes mad because of the "southern strain" of his "heritage." His demise allows her, as she should, to give herself to the proud and healthy son of Kent farmers.[99] Sackville-West provided an even gloomier account of tainted inheritance in *Dragon in Shallow Waters* (1921), an unremittingly brutal account of a blind man who, because he was born bad, murderously abuses both his animals and kin. Nevertheless she ends the book with a stirring depiction of the beauties of the countryside.[100] In these novels, the alarm is raised that the forces of sex, heredity, and reproduction, if not rationally channeled, can have disastrous consequences. And whereas the teeming city and the ugly factory produce pathological degenerates, she presented the countryside as providing not just a geographical but a spiritual retreat.[101]

In 1936, Baker contributed a chapter on "Tree and Shrub Life" to Henry Williamson's *Nature in Britain*. Williamson was one of Britain's most famous naturalists, best known for his animal books *Tarka the Otter* (1927) and *Salar the Salmon* (1935). He was an early environmentalist who found in nature a clean, healthy world produced by the ruthless Darwinian struggle for survival. He contrasted that endangered world to the festering modern urban Britain represented by "big business, fornication, and death . . . chromium fittings, radio, love with pessary, rubber girdles, perms, BBC gentility and the sterilising of truth, cubic international-type concrete architecture . . . white sepulchral bread, gin, and homosexual jokes in the Shaftesbury Avenue theatres . . . world leadership and freedom from tradition. . . . Hoardings, brittle houses, flashiness posing as beauty, mongrel living and cosmopolitan modernism, no planning, all higgledy-piggledy."[102] Such views led him to admire Oswald Mosley, support the British Union of Fascists, and attend the Nuremberg rally. His hailing of Hitler as the "light-bringing phoenix" was to lead during the war to his interrogation, but he managed to avoid internment.[103]

Few Men of the Trees were avowed fascists, but in hankering nostalgically after a lost mythical rural countryside governed by paternalistic landowners, Baker voiced the views of many in the 1930s that there was something wrong about modern mass politics. "Universal franchise has obvious and serious defects when reviewed by thoughtful people, especially when it is extended to those who are unproductive and depending entirely upon the government for support. . . . I do not believe the aristocracy is dead or dying; that would be pessimistic. I believe we need a new aristocracy—an aristocracy of mind and understanding capable, like foresters, of taking a long view of world affairs."[104] For Baker, deforestation was a tragic symptom of the old landowning class's loss of power. As he imagined it, in the past farsighted landowners planted trees knowing that in a century or so their heirs would benefit from them. But whereas forests were once threatened by charcoal burners, in a democratic age they were threatened by death duties. Such swinging taxes removed the incentive of landlords to plant trees, led to the breaking up of fine old estates, pushed woodlands onto the commercial market, and opened the countryside to the intrusion of urban speculators.[105] In a democratic age, big landowners could do little to protect trees, and therefore the challenge had to be taken up by organizations such as the Men of the Trees.

Baker's Men of the Trees movement was part of the climaxing in England of a nostalgia for rural life.[106] There were, of course, some naysayers. Perhaps the most successful was Stella Gibbons, whose novel *Cold Comfort Farm* (1932) hilariously portrayed country living as wet, cold, and filthy. The urbane heroine regards her kin as dim as their four milk cows—Feckless, Graceless, Pointless, and Aimless—and is equally unimpressed by the joys of nature: "A bird sang his idiotic recitative from the dairy roof."[107] Even a conservative writer like Evelyn Waugh depicted an urbanite's unease while in the country. "Normal life, as he saw it, consisted in regular journeys by electric train, monthly cheques, communal amusements and a cosy horizon of slates and chimneys; there was something unEnglish and not quite right about 'the country,' with its solitude and self-sufficiency, its bloody recreations, its darkness and silence and sudden, inexplicable noises; the kind of place where you never know from one minute to the next that you might be tossed by a bull or pitch-forked by a yokel or rolled over and broken by a pack of hounds."[108]

But these comic novels did little to stem the flood tide of books that appeared between 1930 and 1945 praising the joys of farms and villages.[109] H. V. Morton's *In Search of England* (1927)—meaning the land of untouched rural villages—by 1943 had sold something like a million copies.[110] Another classic bewailing the loss of the countryside was *Britain and the Beast* (1938), the "beast" being urban blight.[111] Such books tended to provide comfortable and reassuring views of a past uncontaminated by city sprawl. Such nostalgia was the surest sign that the present was found wanting. The causes for current unease and insecurities included landowners' loss of power, fear of laissez-faire capitalism, and the decline or industrialization of agriculture. Most authors were not opposed to every aspect of modernization and hoped to ally preservation and progress. Baker could probably be best located in this current, but he had a number of far more conservative friends who defended traditional authority against progressive experts.

High Tory types called for the reestablishment of great estates and a new peasantry. They of course agreed with Baker in holding that urban progress led to deforestation, that forests could not be separated from fields, and that death duties "robbed" the landowner.[112] One is reminded that in D. H. Lawrence's *Lady Chatterley's Lover* Clifford has sacrificed his estate's trees during the war but now sees the need to preserve the remainder against the "new England."[113] Strikingly enough some friends of Baker combined an interest in organic farming and attacks on artificial fertilizers with the hopes of reestablishing peasant holdings. Their dream was for the reestablishment of a decentralized society of small, self-supporting villages. They were drawn to guild socialism and Social Credit, and open in their distaste for democracy, which they held was leading toward a servile state. Taking it as a given that some were born to lead and others to follow, they asserted that such policies as mass education, in raising unrealistic ambitions, were creating a generation of misfits.[114] Their targeting of "usury," "marketeers," and agitators revealed a half-concealed anti-Semitism.[115]

Hugh J. Massingham led the Kinship in Husbandry, a sort of landowners' think tank, which had as its goal the regeneration of the countryside.[116] In his stress on organic farming and the peasant holding, Massingham linked soil health, genetics, blood, and the body. His underlying notion was that the English race was linked to the soil.[117] These sorts of ruralists regarded the growing power of the urban working class with fear and loathing. In 1940, Major General W. P. H. Hill wrote Baker of the necessity of inculcating a respect for the countryside. "Although there may be little we can do during the war I am quite certain that this should be one of the leading subjects to consider when the war is over. Unless we have some scheme prepared for dealing with unemployment the country may relapse into a state of anarchy."[118] For Massingham, the Labour Party's election victory in 1945 marked "a bound forward to despotism unknown since the dictatorships of Henry VIII and Cromwell."[119]

Flirting with such ultraright types were some of the pioneers of the organic farming movement. They included Sir George Stapledon, professor of agricultural botany at the University College of Wales. Stapledon asserted that humans were agriculture's most important product. In the country, genes were less mixed, and as a result one had a "sounder" or purer stock. The countryman was certainly superior to the city dweller. "The British countryside in short carries in its population the genes, unsullied and uncontaminated, that maintain and perpetuate our national vigour and our national characteristics."[120] He later took the opportunity, when discussing Disraeli's notion of race, to argue that while complete racial purity was impossible to achieve, "reservoirs of pure race are a valuable heritage" with which to rejuvenate "mixed stocks." He went on to warn that "a general indiscriminate hybridization between the races of *homo sapiens,* if carried to extreme lengths and over most of the globe, might be disastrous. Since world intercourse is sure to enter upon a new phase of intensity after the war, this is a matter that demands rigorous scientific investigation."[121]

Another associate of the Kinship in Husbandry was Sir Albert Howard, the founder in England of the organic farming movement. In India, he had learned of the importance of composting, which linked the health of humans, plants, and animals. He employed a eugenic language, condemning artificial fertilizers for allowing the survival of "weakly plants" that should have perished.[122] Turning to humans, he argued in *The Maintenance of Soil Fertility* (1945) that a high birth rate was a sign of a healthy population, but urbanization, industrialization, and the use of artificial fertilizers depleted the fertility of soil and consequently the fertility of the human population.[123] Howard opposed the mechanization of agriculture because it deprived the soil of nutrients. "The slaves of the Roman Empire have been replaced by mechanical slaves. The replacement of the horse and the ox by the internal combustion engine and the electric motor is, however, attended by one great disadvantage. These machines do not void urine and dung and so contribute nothing to the maintenance of soil fertility. In this sense the slaves of Western agriculture are less efficient than those of ancient Rome."[124] The same argument was made in the journal of Men of the Trees, in which the author lamented, "we are becoming a robot-ridden race."[125]

A progressive such as J. B. S. Haldane who was happy to see the decline of agricultural labor could protest that the factory worker "seems to me a higher type of person from most points of view."[126] It was far more common for interwar commentators to contrast healthy peasants relying on balanced diets with degenerate urban dwellers addicted to processed foods.[127] The countryside, its self-appointed defenders claimed, provided a racial redoubt. Pierse Loftus asserted that agriculture was the nation's most important industry. It produced healthy citizens whereas the city harbored the unfit, the exhausted, and the alien.[128] Laurence Easterbrook concluded his pamphlet by saying of rich soil, "We look to it to breed men and women of character and skill as the most important of all that it produces."[129] The historian G. M. Trevelyan lamented that of all European countries England was the most urbanized and so most threatened with "breeding a race apart from nature."[130] G. T. Wrench went so far as to claim that because the British were cut off from the soil, "We are to-day no longer whole or healthy physically or mentally. In the careful work of the Peckham investigators it has been established that the vast majority of us are subnormal."[131]

In 1936, Baker established a journal entitled *Trees* and in 1938 organized the first Men of the Trees summer school in Oxford. Participants included Rolf Gardiner.[132] Gardiner, a Dorset farmer, called for large estates to serve as centers for rural reconstruction and was involved in the English Mistery, a shadowy back-to-the-land movement made up of monarchists and ruralists. His interest in all things rural was reflected in his support of folk dancing and organic farming.[133] More disturbing were his antiurban, anti-Semitic, and Nordic racist leanings.[134] Such conservatives, impatient with liberal democracy, often cited the empire as providing a better model of administrative efficiency. It is noteworthy that a number had ties to the colonies. Baker worked in Kenya and Nigeria, Rolf Gardiner's family had estates in Malawi, Lymington had land in Kenya, and Morton retired to South Africa. They pined for a hierarchical society that no longer existed in Britain. They were also impressed in the 1930s by Germany's and Italy's purported attempts to protect rural life. It was typically argued that despite what one thought of the Nazis, one had to admit that Germany had gone the furthest in passing legislation for the protection of the countryside.[135]

Pro-German sentiments were particularly prominent in the works of Viscount Lymington (one-time Conservative MP and later Earl of Portsmouth).[136] Lymington, an old friend of Baker, agreed that forests taught both patience and the power of heredity. Turning to humans, he declared that mental and spiritual characteristics were inherited. The interbreeding of races thus posed dangers.[137] The decline in the English birth rate he regarded as a symptom of the neglect of the soil and thus human and spiritual health. Bad food made bad bodies. Taking a crude eugenist line, Lymington asserted that the English state exacerbated the population problem. Mass education and unemployment insurance acts created subnormals who subsisted on the dole. The fall of the infant mortality rate meant that babies were kept alive who should have died, and supporting them inflicted a tax on the fit.[138] Similarly women who insisted on working in factories and postponing childbirth had dangerous births that further burdened the medical system. The resulting decline in "our best human stocks" and the increase in the poorest could only mean disaster.[139] "Subnormality," Lymington declared, was now the average. Workers were mere wage slaves, "international usury" had destroyed craftsmanship, and a "cosmopolitan hierarchy" dominated popular entertainment. Were the British, he asked, to follow Germany in returning to wholesome farm produce, or would they give in to the cheap appeal of tin foods, cinemas, and cigarettes? The war finally put a brake on the explicit expression of such pro-fascist sentiments. Henry Williamson was questioned; Captain Thomas Guillaume St. Barbe Baker (Baker's brother) and Captain George Henry Pitt-Rivers were interned.[140]

Before concluding, it is important to note that those preoccupied with protecting nature were not all located on the right. There was a significant degree of concern over the environment and the preservation of the countryside in progressive circles and much negativity about suburban sprawl. At the beginning of the century, Edward Carpenter—countering Fabian fondness for state economic planning and centralization—linked up environmentalism with socialism, anarchism, sex reform, female emancipation, nudism, and animal rights. Though caricatured as a sandal-making vegetarian, Carpenter's concern for afforestation and the rights of working-class ramblers reflected a popular anti–capitalist/large landowner discourse about the countryside.[141] Two other figures mentioned in earlier chapters warrant attention. Dora Russell was an early convert to what we would call environmental issues and became a campaigner in Cornwall after the Second World War. Her work in education and her general animus to the "machine age" was indicative of a belief in the spiritual or sensual value of the rural.[142] Similarly Naomi Mitchison's historical novels were imbued with a quasi-spiritual, often mythic approach to the natural world, and she became, famously, a noted denizen of rural/coastal Scotland in the 1930s.[143]

The supporters of Men of the Trees tended to be quite different folk. Its emergence can be best understood as part of that heady brew of environmentalism, anticapitalism, eugenics, racialism, and rural values that percolated in parts of 1930s Britain. Not infrequently those who wrote on such subjects critiqued each other. They often had as many differences as they had agreements. What united them was their taking for granted the notion of what one might call a *moral geography* or *sexual landscape*. Good human stock existed in the countryside; abnormals proliferated in the city.[144] Today's environmentalists who support social justice issues are poles apart from the interwar conservationists who were fixated on the purity of nature and the purity of gene pool. But in the 1920s and 1930s, it was hardly surprising that the concerns of eugenicists and environmentalists should have overlapped, especially those frightened by threats to existing class and race hierarchies. Both movements were reactions to the perceived dangers of urbanization and industrialization. Both were alarmed by what they took to be the degeneration of the planet's physical and human resources. And both were modern ways of looking at social problems from a biological point of view.[145]

The issue of reproduction was a central preoccupation of such writers. Lymington lumped together all that he hated in the modern world in asking, "Are we condemned for ever by our own inertia to live like plants on trays in the chemical solution of a far from 'brave new world'? Concrete and chromium, Communism, drugs and movement by perpetual explosion in the cylinder are not the natural lot of man any more than is the degeneration of the tremendous creative forces of sex to the mean and loveless licence of the contraceptive."[146] Such condemnations of birth control by ruralists were not uncommon. The contraceptive symbolized the modern world they opposed. Some referred to council flats as "Birth Control Barracks."[147] Stella Gibbons satirized such views in *Cold Comfort Farm*. The heroine meets initial resistance when teaching the hired girl how she can avoid getting pregnant every year.

> "T'es wickedness! T'es flying in the face of Nature!" she burst out fearfully at last.
>
> "Nonsense!" said Flora. "Nature is all very well in her place, but she must not be allowed to make things untidy."

Though the girl's mother agrees it is immoral, she is hardheaded enough to conclude, "All the same, it might be worth tryin'."[148]

It was this sort of shift in thinking that many ruralists sought to counter. Their organic view was that the issues of population and agriculture could not be separated. H. J. Massingham felt that people had been led astray by materialistic appeals but some had "become aware of the fundamental realities which the tin-opener, the wireless, the motor-car and the movie had hidden from them." He complained that even the churches had failed to see the importance of the environment and one found them contradicting themselves in "preaching against contraceptives and accepting the white bread that causes sterility."[149] Most ruralists argued that the countryside bred sound stock whereas the city produced misfits.[150] The more interesting argument made by some was that urban birth control practices were changing the countryside. "The pen of Dr. Stopes waved over the cities, and lambs were not born on the Downs."[151] What the ecologist Michael Graham meant was that smaller families wanted fewer legs of mutton, leading to less sheep producing less dung resulting in an impoverished soil producing less wheat. Innocent consumers were ignorant of the effect their sexual self-interested decisions had on the living earth.

It would be wrong, however, to see the Men of the Trees as opposed to every aspect of modernity. They only opposed what they regarded as its degenerative aspects. They did not simply want to restore an old society; they campaigned for a new world in which their voices would be heard. They were not opposed to planning; indeed they wanted the sort of legislation that would restrict the breeding of the unfit and encourage that of the healthy, that would support the large landowner and restrain the urban speculator. The national park, the reserve, and the concept of the wilderness itself were all products of modernity requiring the intervention and supervision of trained experts like the forester. As in Germany and Italy, a school of British conservatives presented themselves as embracing a variant of modernism that was purged of its decadent associations, in particular the changing sex and gender relations associated with secular urbanism.[152] The goal was the healthy reproduction of trees and humans.

CONCLUSION

It would be misleading to end this study by leaving the impression that only conservative writers were alarmed by the impact modernization was having on population and the environment. Many on the left were equally concerned.[1] George Orwell warrants special attention. Best known for his critique of totalitarianism, he frequently played off the image of the fruitful countryside against the sterile city. As a moralist, he was appalled by mechanized society's tendency to incite the worst hedonistic instincts in humans and its indifference to the desecration of nature. In *Coming Up for Air* (1939), he has his central character castigate a world in which the goal is to "streamline" everything—from office furniture, to bullets, to men. His fondest memories are of his rural English childhood. Modern cities he describes as littered with shoddy products made from artificial materials—rubber, plastic, and glass; in the countryside, the ruined fields are now covered by jerry-built houses, the fishing pools choked with tin cans.[2] In *Nineteen Eighty-Four,* the twentieth century's most famous dystopia, Orwell similarly accompanied his attack on totalitarianism with a nostalgia for Britain's pastoral past. Tellingly, the first time Winston makes love to Julia is in the countryside, which is sharply contrasted with the shabby city. They find refuge in a clearing on the edge of a hazel wood where a thrush sings. Winston is dazzled by the view of fields, footpaths, molehills and old pastures, hedges and elms and knows that nearby is a stream and green pools where fish swim. This "Golden Country" of which he has dreamed mirrors much the same Arcadian vision conjured up by the ruralists we examined in the preceding chapter.[3]

Before turning to Orwell's views on reproduction, it should be noted that in the 1930s a number of fiction writers had preceded him in raising the possibility of the future imposition of a totalitarian regime in Britain.[4] In James Leslie Mitchell's *Gay Hunter* (1934), an archeologist dreams of a ravaged post–atomic war Britain in which Sub-Men revolt against the Hierarchies and fascists continue to make trouble.[5] Joseph O'Neill's *Land under England* (1935) is in many ways true to the traditional underworld romance in which the narrator discovers a hidden land, but by dwelling on the uses of hypnotism to destroy individualism and render the masses obedient, he draws a comparison to Nazi Germany.[6] John Palmer set the scene for *Hesperides: A Looking Glass Fugue* (1936) in a future society where the narrator is told all mating is controlled by the state. "Eugenics. No more family life . . . mating compounds. Impersonal. Controlled experiments. It's a foolish father knows his own child. All in the code."[7] The novel is reminiscent of Huxley's *Brave New World* in its portrayal of a graded society where private desires are discouraged and sex turned to public service.

An explicit portrayal of a right-wing revolution in Britain can be found in Naomi Mitchison's *We Have Been Warned* (1935), which concludes with the fascists about to take power. Censored because of its defense of birth control, sexual experimentation, and the Soviet Union, the novel had limited circulation.[8] *In the Second Year* (1936), Storm Jameson pushes even further the notion of a fascist National State Party emerging in Britain. In her account, Jews lose their jobs and young people are freed of parental restraints, while the ideals of "sexual purity and social sympathy" disappear. And just as a few years earlier Hitler had first employed Röhm to seize power and then purged him, Jameson imagines her British dictator also seeking the assistance of a charismatic homosexual, who attracts "intelligent women, very young men, and men who had the worser part of women in them."[9] In Andrew Marvell's *Minimum Man; or, Time to Be Gone* (1938), a similar fascist movement called the Party of New Freedom in 1950 launches a revolt against a Labour government.[10]

After the Second World War broke out, Rex Warner's *The Aerodrome* (1941) was published. It tells the story of a fascist air force that takes over an English village. The air vice-marshal's goal is to put an end to muddle and inefficiency. He also opposes the notions of parenthood, ownership, locality, and marriage. Young airmen like the narrator are drawn by the radical call to free themselves of moral constraints. Though parenting is frowned on by the new regime, intercourse is encouraged. Women are to be used; men are never to be the "giver." Only when the air vice-marshal advises the narrator to abort his pregnant lover does the young man finally see through this immoral rhetoric. "I felt the full inhumanity of the organization which he had constructed with such an expense of will and which seemed to me now, not only in this instance, to be designed to stifle life which, however misused, was richer in everything but determination than our order."[11]

In implicitly condemning homosexuality and abortion as somehow fostered by a future fascist state, Jameson and Warner echoed the old argument that modernization undermined healthy reproduction. A feminist like Katharine Burdekin advanced a more original and depressing scenario. What would Europe be like after seven centuries of Nazi rule? Burdekin provides an answer in *Swastika Night* (1937). In her bleak patriarchal dystopia, women are kept in cages, boys are removed from their mothers at eighteen months, and homosexual attachments encouraged. If men treat women as soulless pieces of property, deny them the right to refuse sex, and subject them to rape, it is because males fear female procreative power. This planned "Reduction of Women" leads to a demographic imbalance, however, as fewer girl babies are born, and the novel ends with a glimmer of hope as one father recognizes the need for change.[12] In her earlier futurist novel *Proud Man* (1934), Burdekin provided a thorough critique of twentieth-century British sexual practices. The narrator, a brown-skinned, androgynous, self-fertilizing being, finds England to be a country where the inhabitants are constantly preoccupied by both war and sex. They "multiply heedlessly" yet are guilt ridden since male privilege has distorted the natural relations of the sexes. As the phallus is associated with power, girls grow up feeling inferior and sexually frigid. Some women drift into lesbianism; others—finding that machines have taken their jobs—are driven into marriages that are inevitably marred by adultery and prostitution.[13] Burdekin in short used her futurist account to criticize the gender relations of the present rather than laud some mythical past.

The account of a future dystopia that sounds closest to *Nineteen Eighty-Four* is Cyril Connolly's short story "Year Nine" (1938). The prisoner whose story we read has discovered in a picture gallery degenerate art that reveals that in an earlier age some citizens were actually allowed to go unsterilized and engage in "private love." Now all are forced to participate in "Groupbegettingday." He is ultimately arrested, tortured, and found guilty of being excited by art and "inciting to treason a member of the non-rational (and therefore not responsible for her actions) sex."[14] If the topics Connolly succinctly targets—sterilization, state-sanctioned sex, passive females—reminds one of Huxley, the brutal tone of his tale foreshadows Orwell.

A review of this antitotalitarian literature reveals that Orwell was not quite as original a writer as is commonly claimed. It reminds us as well that futurist writers in the 1930s and 1940s continued to judge societies by their attitudes toward sex and gender. Feminist scholars have noted Orwell's misogyny and fixation on manhood.[15] He repeatedly made clear his opposition to abortion and birth control.[16] In *Keep the Aspidistra Flying* (1936), he set the key sex scene during a day in the country. The woman's refusal to make love without a condom launches the hero into a long rant against the crass precautions one must make in a materialistic society.[17] When they later do have sex in a filthy bedsit, she gets pregnant and takes pills. Like Rex Warner, Orwell portrays recourse to abortion as the most damning example of materialism trumping life. In *The Road to Wigan Pier* (1937), he famously lumped "birth control fanatics" in with vegetarians, nudists, and feminists as those who had brought socialism into disrepute.

> We have reached a stage when the very word "Socialism" calls up, on the one hand, a picture of aeroplanes, tractors, and huge glittering factories of glass and concrete; on the other, a picture of vegetarians with wilting beards, of Bolshevik commissars (half gangster, half gramophone), of earnest ladies in sandals, shock-headed Marxists chewing polysyllables, escaped Quakers, birth-control fanatics, and Labour Party backstairs-crawlers. Socialism, at least in this island, does not smell any longer of revolution and the overthrow of tyrants; it smells of crankishness, machine-worship, and the stupid cult of Russia. Unless you can remove that smell, and very rapidly, Fascism may win.[18]

Orwell held the view that given the pursuit of self-interest there was the real danger that in seventy years time Britain's population would be reduced to eleven million, half of whom would be pensioners.[19] In "The English People" (1944), he called for more babies to prove English vitality. Why had family size declined? "No doubt the dearth of babies is partly due to the competing attraction of cars and radios," he replied, "but its main cause is a typically English mixture of snobbishness and altruism." He called for tax reforms. "Any government, by a few strokes of the pen, could make childlessness as unbearable an economic burden as the big family is now." At the same time, he hoped a more child-friendly culture would emerge. Showing no empathy for desperate women, he declared himself appalled "that abortion, theoretically illegal, should be looked on as a peccadillo, and that the main aim of commercial advertising should be to popularize the idea of 'having a good time' and staying young as long as possible."[20] Though hostile to the Eugenics Society because of its social conservatism, Orwell matched it in his suspicion of allowing women control of their fertility.

In *Nineteen Eighty-Four,* Orwell—like so many other futurist writers—presents the state's policing of sexuality as the key expression of its power.[21] In a society where loyalty to family is incompatible with state service, children spy on their parents, and youths join the "Junior Anti-Sex League," joyful love is a form of rebellion. The party does not mind debauchery "as long as it was furtive and joyless." Its goal is to "remove all pleasure from the sexual act." The party does not permit those attracted to each other to marry; the only purpose of marriage is to produce children. "Sexual intercourse was to be looked on as a slightly disgusting minor operation, like having an enema." The state's half serious hope is for all children to be produced by artificial insemination. The final goal is more ambitious. As O'Brien explains it, "But in the future there will be no wives and no friends. Children will be taken from their mothers at birth, as one takes eggs from a hen. The sex instinct will be eradicated. Procreation will be an annual formality like the renewal of a ration card. We shall abolish the orgasm. Our neurologists are at work upon it now."[22] Whereas Huxley foresaw the state encouraging sexual activity, Orwell assumed the state would seek to repress it.[23] Nevertheless their two classic portrayals of the future were in agreement in warning against the horrors of reproduction by design.

The Second World War in many ways ended one cultural era and began another. Wells, Huxley, and Orwell are today only the best known of that small army of early twentieth-century British writers who pondered the impact that technology, urbanization, science, sex reform, and feminism would have on public and private life. Those who tackled such problems inevitably came round to the question of how such changes would affect sex and reproduction. Many responded to the challenges of a modern, technological world by arguing that sexuality should be brought up to date. Others were horrified by such notions. Eugenic ideas colored the responses of many of those worried by a series of early twentieth-century interrelated sexual crises. Would modernization assure or undermine social stability, gender order, and racial health? Were women having too many or too few children? Were increasing numbers of men plagued by impotence? Was homosexuality spreading? Eugenics offered its answers, inciting the sexual activity of some while constraining that of others. Many of the same people who defended doctors' duty to sterilize the unfit, lauded those who offered hormone treatments to sexually rejuvenate the fit.

The British used the justification of eugenic concerns to discuss sex far more openly than is commonly believed. Several recent accounts of sexuality in early twentieth-century Britain have stressed the sexual ignorance of the mass of the population, as revealed by their failure to employ modern contraceptives.[24] But when one looks at middlebrow culture, one finds no shortage of references to sexuality. Such references necessarily occurred in marriage manuals and birth control tracts. They obviously figured in the literature devoted to more technical subjects such as hormone treatments and the possibility of artificial insemination. Even accounts of motorcars and robots could be given a sexual gloss. Especially significant was the outpouring of books that pondered how sexual and family relations would evolve in the future. Almost anyone writing about modernity—capitalism, mass society, urbanization, technology—and its key icons felt obliged to say something about sex and its eugenic import. For the purposes of this study, no time has been wasted in determining whether a text should be described as science fiction or scientific romance or speculative fiction. Arbitrary literary barriers have been broached by positing that novelists, eugenicists, birth controllers, demographers, doctors, conservationists, and indeed anyone talking of sex and reproduction in the future, were in effect producing "science fiction."

A review of their discussions reminds us of the liveliness with which several generations of writers tackled a host of questions precipitated by advances in medicine and the biological sciences on the one hand and women's attempts to expand their freedoms on the other. It also reveals that many of the concerns we have today about apparently disparate issues ranging from reproductive technologies to protection of the environment can better be understood when traced back to the early twentieth-century British debates over the subjection of human life to scientific management. A common preoccupation with the potentially negative consequences of modernization linked a number of movements. As we have seen, many scientific theories and notions of the future, especially the forebodings, were in fact crystallizations of current social concerns.

Why did such a literature, such an "archive of the future," emerge when it did? It is no surprise that the discussion of the future in the 1920s and 1930s was so spirited. A generation that had survived four years of bloodletting naturally hoped that the page had been turned on the past. And why did an apparent decline in such texts occur in the 1940s? It might be that in battling Nazi Germany and Fascist Italy—two regimes explicitly advancing their own futurist agendas—British writers recoiled from the cruel specter of the actual disciplining of bodies, of eugenics in practice. But as we have noted, eugenic arguments continued to be brandished well after 1945. What is so remarkable, however, is that for several decades Britain produced an enormous number of books spelling out how, in the future, sex and the race might be manipulated. Despite the existence of such an imposing literature, we still know relatively little about how concerns about reproduction permeated the general culture. Some sense is given by broaching the question in a roundabout fashion, tracing the ways in which cultural anxieties about apparently dissimilar subjects—futurist novels, motorcars, robots, hormones, test-tube babies, and trees—were linked to worries about reproduction. In unpacking the complicated cultural resonances of such concerns, we are provided with both a better understanding of interwar British society and an appreciation of the many facets of today's debates over reproduction.

NOTES

INTRODUCTION

1. On British modernity, see Tim Armstrong, *Modernism, Technology and the Body: A Cultural Study* (Cambridge: Cambridge University Press, 1998); Nicholas Daly, *Literature, Technology, and Modernity, 1860–2000* (Cambridge: Cambridge University Press, 2004); David Bradshaw, ed., *A Concise Companion to Modernism* (Oxford: Blackwell, 2003); and see also Mark Selzer, *Bodies and Machines* (New York: Routledge, 1992).

2. On their marriage, see Angus McLaren, *Impotence: A Cultural History* (Chicago: University of Chicago Press, 2007), 167–69.

3. For different approaches to other nation's views of the future, see Joseph J. Corn and Brian Horrigan, *Yesterday's Tomorrows: Past Visions of the American Future* (Baltimore: Johns Hopkins University Press, 1996); Roxanne Panchasi, *Future Tense: The Culture of Anticipation in France between the World Wars* (Ithaca, NY: Cornell University Press, 2009).

4. This study focuses on many of the anxieties of the elite. Because popular culture was forging new kinds of heterosexual and homosexual spaces, working-class people and, in particular, young working-class men and women, experienced modernity, technology, and sexuality in a less apprehensive way. A number of historians have underlined the unevenness in which in gendered and class terms modernity, technology and sexuality were experienced. See, for example, Ross McKibbin, *Classes and Cultures* (Oxford: Oxford University Press, 1998); Billie Melman, *Women and the Popular Imagination in the Twenties: Flappers and Nymphs* (Basingstoke: Macmillan, 1988); Matt Houlbrook, *Queer London: Per-*

ils and Pleasures in the Sexual Metropolis, 1918–1957 (Chicago: University of Chicago Press, 2006).

5. See Jon Turney, *Frankenstein's Footsteps: Science, Genetics and Popular Culture* (New Haven, CT: Yale University Press, 1998).

6. See Susan Merrill Squier, *Babies in Bottles: Twentieth-Century Visions of Reproductive Technology* (New Brunswick, NJ: Rutgers University Press, 1994).

7. Marshall Berman, *All That Is Solid Melts into Air* (London: Verso, 2000), 13.

8. G. R. Searle, *Eugenics and Politics in Britain, 1900–1914* (Leyden: Noordhoff International, 1976); P. M. Mazumdar, *Eugenics, Human Genetics and Human Failings: The Eugenics Movements, Its Sources and Its Critics in Britain* (London: Routledge, 1992); Elazar Barkan, *The Retreat of Scientific Racism: Changing Concepts of Race in Britain and the United States between the World Wars* (Cambridge: Cambridge University Press, 1992); Richard A. Soloway, *Demography and Degeneration: Eugenics and the Declining Birthrate in Twentieth-Century Britain* (Durham: University of North Carolina Press, 1995); Dan Stone, *Breeding Superman: Nietzsche, Race and Eugenics in Edwardian and Interwar Britain* (Liverpool: Liverpool University Press, 2002).

9. On the United States, see Wendy Kline, *Building a Better Race: Gender, Sexuality, and Eugenics from the Turn of the Century to the Baby Boom* (Berkeley: University of California Press, 2001).

10. Zygmunt Bauman, *Modernity and the Holocaust* (Cambridge: Polity Press, 1989), 70.

11. Jeffrey Weeks, *Sex, Politics and Society: The Regulation of Sexuality since 1800* (London: Longman, 1989); Lesley A. Hall, *Sex, Gender and Social Change in Britain since 1880* (London: St. Martin's Press, 2000).

12. Aldous Huxley, "Crebillon the Younger" (1936), in *Complete Essays*, ed. Robert S. Baker and James Sexton (Chicago: Ivan R. Dee, 2000), 4:62–63.

CHAPTER 1

1. Aldous Huxley, *Brave New World* (New York: HarperPerrenial, 1998 [1932]), 7.

2. Cyril Connolly, *Enemies of Promise* (London: Penguin Books, 1961 [1938]); David Daiches, *The Novel and the Modern World* (Chicago: University of Chicago Press, 1939), 188–210; George Woodcock, *Dawn and the Darkest Hour: A Study of Aldous Huxley* (New York: Viking Press, 1972); June Deery, *Aldous Huxley and the Mysticism of Science* (London: Macmillan, 1996); David King Dunaway, "Huxley and Human Cloning: Brave New World in the Twenty-First Century," *Aldous Huxley Annual* 2 (2002): 165–79.

3. *Aldous Huxley: The Critical Heritage* (London: Routledge and Kegan Paul, 1975).

4. I. F. Clarke, *The Pattern of Expectation, 1644–2001* (London: Jonathan Cape, 1979).

5. American culture drew a sharper demarcation between literature and science fiction; in Britain writers, such as Huxley and Wells blurred the boundary. The British moreover interested themselves to a greater degree in the biological sciences while American futurists focused more on the physical sciences, as in the possibility of rocket ships. See Lisa Lynch, "Strange Germs and Hopeful Monsters: Alexander Laing's 1930s American Biotechnology Tales," *New Literary History* 36, no. 2 (2005): 247–61.

6. Edward Bellamy, *Looking Backward: 2000–1887* (New York: Modern Library, 1951 [1888]), 213.

7. William Morris, *News from Nowhere,* in *Collected Works* (London: Longmans, Green, 1912), 16:62.

8. H. G. Wells, *A Modern Utopia* (Lincoln: University of Nebraska Press, 1967 [1905]), 294.

9. Norman and Jeanne Mackenzie, *The Time Traveller: The Life of H. G. Wells* (London: Weidenfeld and Nicolson, 1973); Patrick Parrinder, *Shadows of the Future: H. G. Wells, Science Fiction and Prophecy* (Liverpool: Liverpool University Press, 1995); Michael Foote, *H.G.: The History of Mr. Wells* (New York: Doubleday, 1995).

10. Godfrey Sweven [John Macmillan Brown], *Limanora: The Island of Progress* (New York: G. P. Putnam's Sons, 1903).

11. Summer Spring [pseud.], *Backwards and Forwards* (London: Henry J. Glaisher, 1905), 60, 67.

12. For a more sophisticated Russian account of such a dystopia, see Evgenii Ivanovich Zamiatin, *We,* trans. Gregory Zilboorg (New York: Dutton, 1959 [1924]). See also Erika Gottlieb, *Dystopian Fiction East and West: Universe of Terror and Trial* (Montreal: McGill-Queen's University Press, 2001); Richard Stites, *Revolutionary Dreams: Utopian Vision and Experimental Life in the Russian Revolution* (New York: Oxford University Press, 1989); M. Keith Booker, *The Dystopian Impulse in Modern Literature* (Westport, CT: Greenwood Press, 1994).

13. Walter Besant, *Revolt of Man* (London: William Blackwood and Sons, 1882).

14. Elliot Evans Mills, *The Decline and Fall of the British Empire* (London: Simpkin, Marshall, Hamilton, Kent, 1905).

15. Allan Reeth, *Legions of the Dawn* (London: Fisher Unwin, 1908); Jesse Wilson, *When the Women Reign, 1930* (London: Arthur Stockwell, 1909); A. C. Fox-Davies, *The Sex Triumphant* (London: George Routledge and Sons, 1909); Anon., *When Woman Rules! A Tale of the First Women's Government* (London: John Long, Limited, 1923).

16. Laurence Housman, *John of Jingalo: The Story of a Monarch in Difficulties* (London: Chapman and Hall, 1912); and see Sandra Stanley Holton, *Suffrage Days: Stories from the Women's Suffrage Movement* (London: Routledge, 1996), 143.

17. Stephen Leacock, "The Fifty-Fifty Sexes: A Fragment from Utopia," in *Afternoons in Utopia* (London: John Lane The Bodley Head, 1932), 179–92.

18. H. L. Samuel, *An Unknown Land* (London: George Allen and Unwin, 1942), 41.

19. For early feminist portrayals of the future possibility of women freeing themselves of male sexual demands, see Jane Hume Clapperton, *Margaret Dunmore; or, A Socialist Home* (London: Swan Sonnenschein, Lowry, 1888); Jane Hume Clapperton, *Scientific Meliorism and the Evolution of Happiness* (London: Kegan Paul, Trench, Trübner, 1885); George Noyes Miller, *The Strike of Sex* (London: Reeves, 1895); Ellis Ethelmer [Elizabeth Wolstenholme Elmy], *Woman Free* (London: Woman's Emancipation Union, 1893); Frances Swiney, *The Bar of Isis: The Law of the Mother* (London: C. W. Daniels, 1907). For feminist utopias where parthenogenesis is employed so only white females are created, see the Americans Mary E. Bradley Lane, *Mizora: A Prophecy*, ed. Jean Pfaelzer (Syracuse, NY: Syracuse Uni-

versity Press, 2000 [1881]); and Charlotte Perkins Gilman, *Herland* (London: Women's Press, 2001 [1915]).

20. Anon.[Cicely Hamilton?], *Star of the Morning: A Chronicle of Karyl the Great and the Revolt of 1920–22* (London: Burleigh, 1906); Cora Minnett, *The Day after Tomorrow* (London: F. V. White, 1911); Ronald Knox, *Memories of the Future, Being Memoirs of the Years 1915–1972 Written in the Year of Grace 1988 by Opal, Lady Porstock* (London: Methuen, 1923); Victoria Cross [Vivian Corey], *Martha Brown, MP: A Girl of Tomorrow* (London: F. Werner Laurie, 1935).

21. Lady Florence Dixie, *Gloriana; or, The Revolution of 1900* (London: Henry, 1890), 137.

22. Elise Kay Gresswell, *When Yvonne Was Dictator* (London: John Heritage, 1935), 32, 203.

23. Edward George Bulwer-Lytton, *The Coming Race* (Middletown, CT: Wesleyan University Press, 2005), 39.

24. Bulwer-Lytton, *The Coming Race*, 119.

25. Bulwer-Lytton, *The Coming Race*, 196.

26. H. G. Wells, *The Island of Dr. Moreau* (London: Heinemann, 1896).

27. Wells, *A Modern Utopia*, 202.

28. Wells portrayed unattached young women complaining that they were regarded by society as "inflammable litter," which feared that "we should all catch fire." H. G. Wells, *Ann Veronica* (London: T. Fisher Unwin, 1925 [1909–10]), 15.

29. Edward Shanks, *The People of the Ruins: A Story of the English Revolution and After* (London: W. Collins, 1920).

30. Cicely Hamilton, *Lest Ye Die* (London: Jonathan Cape, 1922).

31. James Leslie Mitchell [Lewis Grassic Gibbon], *Gay Hunter* (London: William Heinemann, 1934).

32. D. H. Lawrence wrote Huxley: "It seems to me queer that you prefer to present men chiefly—as if you cared for women not so much for what they were in themselves as for what men saw in them. So that after all in your work women seem not to have an existence, save they are the projections of men." Aldous Huxley, "D. H. Lawrence," in *Complete Essays*, ed. Robert S. Baker and James Sexton (Chicago: Ivan R. Dee, 2000), 3:74.

33. The character of Lenina was possibly inspired by the stories of Panteleimon Romanof; see Peter Edgerly Firchow, *The End of Utopia: A Study of Aldous Huxley's Brave New World* (Lewisburg, PA: Bucknell University Press, 1984).

34. Huxley, *Brave New World*, 36–37, 193–94.

35. Dora Russell, *Hypatia; Or, Woman and Knowledge* (London: Kegan Paul, Trench, Trübner, 1925), 35. And see also Dora Russell, *The Right to Be Happy* (London: George Routledge and Sons, 1927). Dora wrote her husband of their open marriage: "I have a silly kind of feeling that you & I are born to set the world right for the coming generation to give them some idea of how men & women can come together & build society instead of having this bitter sex war all the time & those confounded silly jealousies." Ray Monk, *Bertrand Russell, 1921–70: The Ghost of Madness* (London: Jonathan Cape, 2000), 90.

36. John Gloag provided what sounds like a poison-pen portrait of Dora Russell. "She never concealed her love affairs, for when they married they had agreed upon the conven-

tional post-War unfaithfulness pact. Their friends said it was 'rather nineteen nineteen' and laughed. . . . Her three children were shaggy little savages who discussed excrement, sex and obstetrics with gusto whenever they noticed that it embarrassed people who were not completely modernized." John Gloag, *To-Morrow's Yesterday* (London: George Allen and Unwin, 1932), 29–30.

37. C. E. M. Joad, *Thrasymachus; or, The Future of Morals* (London: Kegan Paul, Trench, Trübner, 1926), 46, 66. Joad's book was one of the eighty-six on the future that between 1924 and 1931 Kegan Paul, Trench, Trübner and Company published in their "To-Day and To-Morrow Series," beginning with J. B. S. Haldane's *Daedalus*. See Max Saunders and Brian Hurwitz, "The To-Day and To-Morrow Series and the Popularisation of Science: An Introduction," *Interdisciplinary Science Review*, 34, no. 1 (March 2009): 3–8. On the "boom in futurology" between 1925 and 1935, see Brian Stableford, "Science Fiction between the Wars: 1918–1938," in *Anatomy of Wonder: A Critical Guide to Science Fiction*, ed. Neil Barron (New York: Bowker, 1981), 93.

38. Huxley, *Brave New World*, 213.

39. Huxley, *Brave New World*, 42.

40. Harry M. Geduld, ed., *The Definitive Time Machine* (Bloomington: Indiana University Press, 1987), 44.

41. Edwin Lester Arnold, *Lieut. Gulliver Jones: His Vacation* (London: New English Library, 1976 [1905]), 30, 33, 49.

42. Max Nordau, *Degeneration* (London: W. Heinemann, 1913), 539; Matt Cook, *London and the Culture of Homosexuality, 1885–1914* (Cambridge: Cambridge University Press, 2003), 95.

43. H. G. Wells, *The Shape of Things to Come* (New York: Macmillan, 1936 [1933]), 400. Wells's novel was the basis of one of the few British interwar science-fiction films, *Things to Come* (1936) directed by William Cameron Menzies. On the small number of fantasy films, see John Sedgwick, *Popular Filmgoing in the 1930s: A Choice of Pleasures* (Exeter: University of Exeter Press, 2000), 158.

44. Charlotte Haldane, *Man's World* (London: Chatto and Windus, 1926), 296, 299. And for the claim that in the current decadent age the fashionably thin woman looked "like a powdered degenerate boy" while among men there were more "sexual perversities," see Ronald Campbell Macfie, *Metanthropos; or, The Body of the Future* (London: Kegan Paul, Trench, Trübner, 1928), 80, 81.

45. Olaf Stapledon, *Odd John* (1935), in *Odd John and Sirius* (New York: Dover Publications, 1972), 14, 48–49.

46. C. A. B., "Highbrow Science Fiction," *Amazing Stories* 7, no. 1 (April 1932): 86.

47. Joseph Needham, *Scrutiny*, May 1932, 77.

48. Bertrand Russell, *The Scientific Outlook* (London: George Allen and Unwin, 1931), 266; and on his critique of current sexual mores, see Bertrand Russell, *The Conquest of Happiness* (London: George Allen and Unwin, 1930), 101, 102. On the prediction of a relaxing of sexual taboos, see also Robert Graves, *Lars Porsena; or, The Future of Swearing* (London: Kegan Paul, Trench, Trübner, 1927), 67–71.

49. H. G. Wells, *Anticipations: Of the Reaction of Mechanical and Scientific Progress upon*

Human Life and Thought (London: Harper and Brothers, 1902 [1901]), 239; and for a supporter, see George Whitehead, *Birth Control and Race Culture: The Social Aspects of Sex* (London: Pioneer Press, 1925).

50. Sweven, *Limanora: The Island of Progress*, 579.

51. Gloag, *To-Morrow's Yesterday*, 143. The opposite account, that humans—because they shrank from childbearing—were ultimately replaced by another species, is told in John Lionel Tayler, *The Last of My Race: A Dream of the Future* (London: Simpkin Marshall, 1924).

52. The device of an aircraft to enter a simpler world is also employed in J. Leslie Mitchell, *Three Go Back* (Indianapolis: Bobbs-Merrill, 1932).

53. James Hilton, *Lost Horizon* (London: Macmillan, 1933), 182.

54. Hilton, *Lost Horizon,* 87, 153, 228. Similarly in Middle Earth, as described by J. R. R. Tolkien's *Lord of the Rings* (London: George Allen and Unwin, 1954–55), there is no sex; it is a mythical schoolboy land fortunately free of women.

55. Marie Stopes, *The First Five Thousand, Being the First Report of the First Birth Control Clinic in the British Empire* (London: Bale and Danielsson, 1925).

56. Geduld, ed., *The Definitive Time Machine,* 49.

57. Geduld, ed., *The Definitive Time Machine,* 50.

58. Geduld, ed., *The Definitive Time Machine,* 51, 57.

59. H. G. Wells, *The Dream: A Novel* (New York: Macmillan, 1924), 76.

60. H. G. Wells, *Men Like Gods,* in *The Sleeper Awakes and Men Like Gods* (London: Oldham Press, 1923), 229.

61. Wells, *Men Like Gods,* 249. For a similar fictional portrait of a future "ordered garden," see John Lawrence Hodgson, *The Time Journey of Dr. Barton* (Eggington, Beds.: John Hodgson, 1929).

62. C. P. Blacker, *Birth Control and the State: A Plea and a Forecast* (London: Kegan Paul, Trench, Trübner, 1926), 33, 34, 40–41.

63. Bertrand Russell, *Icarus; or, The Future of Science* (London: Kegan Paul, Trench, Trübner, 1924), 43. On progressives' enthusiasm for planning, see C. E. M. Joad, *Manifesto: Being the Book of the Federation of Progressive Societies and Individuals* (London: George Allen and Unwin, 1934).

64. Aldous Huxley, "Work and Leisure" (1925), in *Complete Essays*, 1:41.

65. Huxley's account has some contradictions. If women lose their ovaries, why are they supplied with contraceptives? Given the availability of oophorectomies and birth control, why the need for enormous abortion centers? And why is there no mention of male contraceptives?

66. Charlotte Burghes Haldane, *Nature*, April 23, 1932, 597–98.

67. Charlotte Burghes Haldane, *Motherhood and Its Enemies* (London: Chatto and Windus, 1927), 97.

68. Aldous Huxley, *Crome Yellow* (London: Chatto and Windus, 1949 [1921]), 45, 295–96; Aldous Huxley, *Those Barren Leaves: A Novel* (London: Chatto and Windus, 1950 [1925]), 201

69. Aldous Huxley, "After the Fireworks," in *Brief Candles* (London: Chatto and Windus, 1930), 201.

70. Blacker, *Birth Control and the State*, 19.

71. H. G. Wells, *The Wife of Sir Isaac Harman* (London: Macmillan, 1914), 342.

72. Haldane, *Motherhood and Its Enemies*.

73. Enid Charles, *The Menace of Under-Population: A Biological Study of the Decline of Population Growth* (London: Watts, 1936). As a feminist, Charles retained her maiden name after her marriage to the biologist Lancelot Hogben.

74. Gloag, *To-Morrow's Yesterday*, 29. Gloag, an architect, referred to the international style as "robot modernism" and preferred craft-based interior design. See Victoria Rosner, *Modernism and the Architecture of Private Life* (New York: Columbia University Press, 2005), 9; John Gloag and John Mansfield, *The House We Ought to Live In* (London: Duckworth, 1923).

75. Sydney Fowler Wright, *Deluge* (Middletown, CT: Wesleyan University Press, 2003 [1927]).

76. "Justice"(1930), in *S. Fowler Wright's Short Stories* (Ludlow: FWB, 1996), 24–36.

77. Muriel Jaeger, *The Question Mark* (London: Hogarth Press, 1926), 60–61.

78. Michael Arlen, *Man's Mortality* (London: Heinemann, 1933), 14.

79. Olaf Stapledon, *Last and First Men: A Story of the Near and Far Future* (London: Millennium, 1999 [1930]), 66; Leslie A. Fiedler, *Olaf Stapledon: A Man Divided* (Oxford: Oxford University Press, 1983), 56, 112.

80. For a 1908 Russian futurist novel about Mars that discusses euthanasia, the emancipation of women, unisex clothes, and gender-free names, and where the terms liaison, affair, marriage, and romance all mean the same thing, see Alexander Bogdanov, *Red Star: The First Bolshevik Utopia*, trans. Charles Rougle and ed. Loren R. Graham and Richard Stites (Bloomington: Indiana University Press 1984).

81. Wells, *Anticipations*, 334.

82. H. G. Wells, *Mankind in the Making* (London: Chapman and Hall, 1903).

83. H. G. Wells, *In the Days of the Comet* (London: Macmillan, 1906), 53.

84. In his own life, as he explained to his understanding wife, Wells separated love from "physical necessity." "I want a healthy woman handy to steady my nerves & leave my mind free for real things." Norman and Jeanne Mackenzie, *The Time Traveller*, 304.

85. On feminist utopian thinking, see also Lesley A. Hall, '"I have never met the normal woman': Stella Browne and the Politics of Womanhood," *Women's History Review* 6 (1997): 157–82.

86. Vera Brittain, *Halcyon; or, The Future of Monogamy* (London: Kegan Paul, Trench, Trübner, 1929), 13, 38, 42, 47, 85, 86, 87, 88, 89. See also Samuel D. Schmalhausen, *Why We Misbehave* (New York: Macaulay, 1928). On Brittain's own marriage and her "occasional contacts," see Vera Brittain, *Testament of Experience* (London: Fontana, 1979 [1957]), 39, 46, 116.

87. Brittain, *Halcyon; or, The Future of Monogamy*, 54, 57, 58. On the sex antagonism of the prewar years followed in the 1920s by hopes of a "great experiment" in reformed

marriage, see Marcus Collins's *Modern Love: An Intimate History of Men and Women in Twentieth-Century Britain* (London: Atlantic Books, 2003).

88. On the "changing patterns of loving" of the time, see the autobiography by J. B. S. Haldane's sister, Naomi Haldane Mitchison, *You Well May Ask: A Memoir, 1920–1940* (London: Flamingo, 1979).

89. Maurice Eden Paul, *Chronos; or, The Future of the Family* (London: Kegan Paul, Trench, Trübner, 1930).

90. Harry Joseph Birnstingl, *Lares et Penates; or, The Home of the Future* (London: Kegan, Paul, Trench, Trübner, 1928), 85–86.

91. Norman Haire, *Hymen; or, The Future of Marriage* (London: Kegan Paul, Trench, Trübner, 1927), 56–57. And see also Amabel Williams-Ellis, *To Tell the Truth . . .* (London: Jonathan Cape, 1933), a satirical reversal of the anti-Soviet travelogue that portrays a Russian visitor shocked to discover that British women's sexuality is still constrained by taboos and prejudices.

92. John B. Watson, "After the Family—What?," in *The New Generation: The Intimate Problems of Modern Parents and Children,* ed. V. F. Calverton and Samuel D. Schmalhausen (London: George Allen and Unwin, 1930), 72.

93. Bertrand Russell, *Marriage and Morals* (London: George Allen, 1929), 164. And see Brian Harrison, "The False Consciousness of a Feminist," in *Intellect and Social Conscience: Essays on Bertrand Russell's Early Work,* ed. Margaret Moran and Carl Spadoni (Hamilton: McMaster University Press, 1984), 157–207.

94. G. E. Newsom, *The New Morality* (London: Ivor Nicholson and Watson, 1932), 256–57, 261. And for a fictional attack on sex reformers, see Leonard Rossiter, *The Sex Age* (London: Selwyn and Blount, 1928).

95. Horace Wykeham Can Newte, *The Master Beast: Being a True Account of the Ruthless Tyranny Inflicted on the British People by Socialism, A.D. 1888–2020* (London: Rebman Limited, 1907), 93, 119, 142, 235. The book was updated as Horace Wykeham Can Newte, *The Red Fury: Britain under Bolshevism* (1919); and see also Everett F. Beiler, ed., *Science-Fiction, the Early Years* (Kent, OH: Kent State University Press, 1990), 542.

96. F[rederick] Britten Austin, *The Red Flag* (London: Eyre and Spottiswoode, 1932), 389. For an account of a future battle against a Labour government and its "Red Terror," see Hugh Addison, *The Battle for London* (London: Herbert Jenkins, 1923).

97. Controlled breeding was also attacked in John Kendall (Margaret Maud Brash), *Unborn To-Morrow* (London: W. Collins, 1933). Herek, a "Grade Two man," is drawn to an ungraded college girl but debarred from marriage due to his TB. The world state moreover supports miscegenation to overcome nationalism. Proud of his Nordic blood Herek escapes to the Lake District to join other rebels.

98. Gavin Schaffer, *Racial Science and British Society, 1930–62* (Basingstoke: Palgrave Macmillan, 2008).

99. Karl Pearson, *Life, Letters and Labours of Francis Galton* (Cambridge: Cambridge University Press, 1930), IIIA:414, 417.

100. Pearson, *Life, Letters and Labours of Francis Galton*, IIIA:422.

101. J. C. David, "Science and Utopia: The History of a Dilemma," in *Nineteen Eighty-Four:*

Science between Utopia and Dystopia, ed. Everett Mendelsohn and Helga Nowotny (Dordrecht: Reidel, 1984), 39.

102. Peter Weingart, "Eugenic Utopias—Blueprints for the Rationalization of Human Evolution," in Mendelsohn and Nowotny, *Nineteen Eighty-Four: Science between Utopia and Dystopia,* 173–87.

103. On continental portrayals of state supervision of marriages, see Paolo Mantegazza, *L'Anno 3000* (Milano: Treves, 1897); Ernest Tarbouriech, *La Cité future: Essai d'une utopie scientifique* (Paris: P-V Stock, 1902), 298–307.

104. A. L. Morton, *The English Utopia* (London: Lawrence and Wishart, 1969); and on eugenics in fiction, see Peter Morton, *"The Busiest Man in England": Grant Allen and the Writing Trade, 1875–1900* (New York: Palgrave Macmillan, 2005); Angelique Richardson, *Love and Eugenics in the Late Nineteenth Century: Rational Reproduction and the New Woman* (Oxford: Oxford University Press, 2003).

105. Edward Bellamy, *Looking Backward: 2000–1887* (New York: Modern Library, 1951 [1888]), 218. And on progressives and eugenics, see Angus McLaren, "Sex Radicalism in the Canadian Pacific North West, 1890–1920," *Journal of the History of Sexuality* 2 (1992): 527–47.

106. Patrick Parrinder, "Eugenics and Utopia: Sexual Selection from Galton to Morris," *Utopian Studies* 8, no. 2 (1997): 1–12.

107. Wells, *A Modern Utopia,* 142.

108. Wells, *Mankind in the Making,* 298–99.

109. Wells, *A Modern Utopia,* 195–96.

110. Lord Birkenhead (Frederick Edwin Smith), *The World in 2030 A.D.* (London: Hodder and Stoughton, 1930).

111. George Godwin, *Cain; or, The Future of Crime* (London: Kegan Paul, Trench, Trübner, 1928), 78.

112. Julian Huxley, "The Applied Science of the Next Hundred Years: Biological and Social Engineering," *Life and Letters* 11 (1934): 38–46.

113. Birkenhead, *The World in 2030 A.D.,* 14; and see also 83.

114. Blacker, *Birth Control and the State,* 85.

115. Wells, *Anticipations,* 314.

116. In *Mankind in the Making* (1903), Wells toned down the racist remarks of *Anticipations* (1901). Nevertheless in *In the Days of the Comet* (1906), he presents a Jew confessing to his race's self-conceit.

117. Birkenhead, *The World in 2030 A.D.,* 178. For the similar argument that social "parasites" were thriving in 1950s Britain and if nothing were done they would eventually kill their host, see Charles Galton Darwin, *The Next Million Years* (London: Rupert-Hart Davis, 1952), 94.

118. Godwin, *Cain; or, The Future of Crime,* 32, 33, 50.

119. F. C. S. Schiller, *Tantalus; or, The Future of Man* (London: Kegan Paul, Trench, Trübner, 1924), 53, 63–64.

120. Haire, *Hymen; or, The Future of Marriage,* 74, 75–76, 91.

121. Frederick le Gros Clarke, *Between Two Men* (London: Boriswood, 1935), 25–28.

122. Vernon Lee [Violet Paget], *Proteus; or, The Future of Intelligence* (London: Kegan Paul, Trench, Trübner, 1925), 58.

123. Sydney Fowler Wright, "PN 40" (1929), in *S. Fowler Wright's Short Stories* (Ludlow: FWB, 1996), 96–120.

124. Russell, *Hypatia; or, Woman and Knowledge*, 60.

125. Charles, *The Menace of Under-Population*, 126–27, 138.

126. Haldane *Motherhood and Its Enemies*, 232.

127. J. B. S. Haldane, "The Place of Science in Western Civilization," *The Realist* 2, no. 2 (1929): 151. On left-wing eugenics, see also Eden Paul, "Birth Control: Communist and Individualist Aspects," *Medical Critic and Guide* 25, no. 6 (June 1922): 212–16.

128. Mark Adams, "'Last Judgment: The Visionary Biology of J. B. S. Haldane," *Journal of the History of Biology* 33 (2000): 480.

129. Russell, *Icarus; or, The Future of Science*, 47.

130. Russell, *Icarus; or, The Future of Science*, 48–49; and see Monk, *Bertrand Russell*, 105.

131. Julian Huxley, *Man Stands Alone* (New York: Harper, 1941), 66. Huxley also co-authored a book aimed at a popular audience that heralded the prospect of "Life under Control." In the future, "sterile jazz-dancers and joy-riders" would not reproduce thanks to a "a definite reproductive policy." See H. G. Wells, Julian Huxley, and G. P. Wells, *Science of Life* (London: Cassell, 1930), 879.

132. John Desmond Bernal, *The World, Flesh, and the Devil* (London: Kegan Paul, Trench, Trübner, 1929), 95.

133. In 1939, the American geneticist Herman Muller called for both negative and positive birth control in the forms of sterilization, abortion, and artificial insemination. He argued that childbearing had to be honored, whether the parents were married or not, but some kind of "selection" was needed if progress was to take place. In more primitive times, such selection had occurred automatically; now guidance or "conscious selection" was needed. Herman Muller, "Social Biology and Population Improvement," *Nature* 144 (September 16, 1939), reprinted in *Man's Future Birthright: Essays on Science and Humanity*, ed. Elof Axel Carlson (Albany: State University of New York Press, 1973), 114. See also Diane Paul, "Genetic Engineering and Eugenics: The Uses of History," in *Is Human Nature Obsolete? Genetics, Bioengineering, and the Future of the Human Condition*, ed. Howard W. Baillie and Timothy K. Casey (Cambridge, MA: MIT Press, 2005), 130–35.

134. Susan Merrill Squier, *Babies in Bottles: Twentieth-Century Visions of Reproductive Technology* (New Brunswick, NJ: Rutgers University Press, 1994).

135. Walter Heape, "The Artificial Insemination of Mammals and Subsequent Possible Fertilisation or Impregnation of Their Ova," *Proceedings of the Royal Society of London*, 1897, 52–63.

136. Philip J. Pauly, *Controlling Life: Jacques Loeb and the Engineering Ideal in Biology* (New York: Oxford University Press, 1987), 93, 100.

137. Andrés Horacio Reggiani, *God's Eugenicist: Alexis Carrel and the Sociobiology of Decline* (New York: Berghahn Books, 2006).

138. Robert Bud, *The Uses of Life: A History of Biotechnology* (Cambridge: Cambridge Uni-

versity Press, 1993), 70–72; and for a defense of the materialistic approach of science, see Joseph Needham, *Man a Machine* (London: Kegan Paul, Trench, Trübner, 1927).

139. J. B. S. Haldane, *Daedalus; or, Science and the Future* (London: Kegan Paul, Trench, Trübner, 1924), 4.

140. Haldane, *Daedalus; or, Science and the Future*, 64.

141. On the application of Haldane's ideas by Gregory Pincus, see Pauly, *Controlling Life: Jacques Loeb and the Engineering Ideal in Biology*, 187–93.

142. Haldane, *Daedalus; or, Science and the Future*, 66–67.

143. Huxley, *Crome Yellow*, 46–47. In another novel, Huxley has a character exclaim that it is a great age to live in: anything may be discovered—"the artificial creation of life, the proof of continued existence after death—why, it's all infinitely exhilarating." Huxley, *Those Barren Leaves*, 34.

144. For a less optimistic view, see Paul, *Chronos; or, The Future of the Family*, 35.

145. E. E. Fournier d'Albe, *Quo Vadimus? Some Glimpses of the Future* (London: Kegan Paul, Trench, Trübner, 1925), 48.

146. Russell, *Icarus; or, The Future of Science*, 10.

147. Bertrand Russell, *Marriage and Morals* (London: George Allen, 1929), 207.

148. Russell, *Marriage and Morals*, 214.

149. Bertrand Russell, *The Scientific Outlook* (London: George Allen and Unwin, 1931), 261–62.

150. Russell, *Icarus; or, The Future of Science*, 50.

151. In 1935, H. J. Muller asserted that "Only social inertia and popular ignorance now hold us back from putting into effect (at least in a limited experimental way) such a severance of the function of reproduction from the personal love-life of the individual." His worry was that under capitalism science would produce hordes of Billy Sundays, Valentinos, Babe Ruths, and Al Capones instead of Lenins, Newtons, and Pasteurs. H. J. Muller, *Out of the Night: A Biologist's View of the Future* (New York: Vanguard Press, 1935), 111.

152. Julian Huxley, "The Tissue Culture King: A Parable of Modern Science" *Yale Review*, April 1926, 479–504.

153. Anthony Ludovici, *Lysistrata; or, Woman's Future and Future Woman* (London: Kegan Paul, Trench, Trübner, 1924), 90.

154. Ludovici, *Lysistrata; or, Woman's Future and Future Woman*, 89.

155. Owen Johnson, *The Coming of the Amazons: A Satiristic Speculation on the Scientific Future of Civilization* (New York: Longmans, Green, 1931), 113, 120. That the whole story is simply a madman's dream is suggested in the novel's conclusion.

156. Birkenhead, *The World in 2030 A.D.*, 176, 178. For a critique of Birkenhead, see Margaret Kornitzer, *The Modern Woman and Herself* (London: Jonathan Cape, 1932), 28–30.

157. Birkenhead, *The World in 2030 A.D.*, 15.

158. Haldane, *Man's World*, 52, 55, 56, 57, 59.

159. Yet elsewhere as in *Motherhood and Its Enemies* (1927) Haldane warns against eugenic "cranks." See also Judith Adamson, *Charlotte Haldane: Woman Writer in a Man's World* (London: Macmillan, 1998).

160. Charles, *The Menace of Under-Population*, 191–92.

161. Brittain, *Halcyon; or, The Future of Monogamy*, 76.

162. Huxley, *Brave New World*, 7; 159–60.

163. Blacker, *Birth Control and the State*, 88. Though eugenicists lamented the purported physical and mental decline of the masses and campaigned for the improvement of the race, there were some like Aldous Huxley who saw that a future society would actually need the unintelligent to do the routine repetitive undesirable work. "A Note on Eugenics" (1927), in Aldous Huxley, *Complete Essays*, ed. Robert S. Baker and James Sexton (Chicago: Ivan R. Dee, 2000), 2:284–85.

164. Joseph Needham, "Notes on the Way," *Time and Tide*, 12 (September 10 1932): 970–72.

165. Birkenhead, *The World in 2030 A.D.*, 16.

166. Laurence Manning, *The Man Who Awoke* (1933) (London: Sphere Books, 1977), 121. Even a sex manual such as Edward Charles's echoed *Brave New World* in its vision of birthing. See Edward Charles [Charles Edward Hempstead], *An Introduction to the Study of the Psychology and Physiology and Bio-Chemistry of the Sexual Impulse* (London: Boriswood, 1935), 200–201.

167. Stapledon, *Odd John*, 139.

168. Fredric Jameson, "Progress versus Utopia; or, Can We Imagine the Future?" *Science Fiction Studies* 27 (July 1982): 153.

169. Woodcock, *Dawn and the Darkest Hour: A Study of Aldous Huxley*, 176.

170. Q. D. Leavis, *Scrutiny* 5 (September 1936): 180.

171. P. H. Houston, "The Salvation of Aldous Huxley," *American Review*, December 1934, 231.

CHAPTER 2

1. Aldous Huxley, *Brave New World* (New York: HarperPerrenial, 1998 [1932]), 48.

2. "The Limitation Of Noise," *The Times*, August 31, 1928, 13; and see E. E. Fournier d'Albe, *Quo Vadimus? Some Glimpses of the Future* (London: Kegan Paul, Trench, Trübner, 1925), 32–33.

3. Allen Samuels, "Accidents: The Car and Literature," in *Autopia: Cars and Culture*, ed. Peter Wollen and Joe Kerr (London: Reaktion Books, 2003), 58. On the mystique of the less ubiquitous airplane, see Martin Francis, *The Flyer: British Culture and the Royal Air Force, 1939–1945* (Oxford: Oxford University Press, 2008)

4. Vance Packard's *The Hidden Persuaders* (New York: D. McKay, 1957) and John Keats's *The Insolent Chariots* (Philadelphia: J. B. Lippincott, 1958) both advanced the now well-known argument that in the post–World War Two period cars were advertised to appeal primarily to buyers' emotional needs, that automobiles were made sexy. It was not by chance that some were phallic shaped while others were touted as being more responsive than any woman to a man's touch.

5. Aldous Huxley, "Wanted: A New Pleasure," in *Complete Essays*, vol. 3, *1930–1935*, ed. Robert S. Baker and James Sexton (Chicago: Ivan R. Dee, 2001), 263; and see also Stephen Bayley, *Sex, Drink and Fast Cars: The Creation and Consumption of Images* (London: Faber

and Faber, 1986); Kristin Ross, *Fast Cars, Clean Bodies: Decolonization and the Reordering of French Culture* (Cambridge, MA: MIT Press, 1995).

6. Osbert Sitwell, *Great Morning* (London: Macmillan, 1948), 234–35. The sporting motorist Captain J. F. C. Kruse noted that car manufacturers tapped into this fixation, only using the term "performance" to refer to a car's speed and economy, not to its safety or durability. "Design and Performance," *The Realist* 2, no. 1 (1929): 110.

7. R. Austin Freeman, *Social Decay and Regeneration* (London: Constable, 1921), 196.

8. Sean O'Connell, *The Car in British Society: Class, Gender and Motoring, 1896–1939* (Manchester: Manchester University Press, 1998), 190. According to the French writer Georges Duhamel, the car made us more brutal than we already were, it banished concern for others. Its speed was needed if only to escape shame. The car did not conquer space—it wasted it. Georges Duhamel, *Scènes de la vie future* (Paris: Mercvre de France, 1956 [ca. 1930]), 85; *America the Menace: Scenes from the Life of the Future*, trans. Charles Miner Thompson (London: George Allen and Unwin, 1930).

9. Anthony M. Ludovici, "Back to the Jungle," *New English Weekly* 32 (1947–48): 145.

10. Anthony M. Ludovici, "Slavery, Life and Revolution," *New English Weekly* 19 (1941): 7–8. See D. Stone, "The Extremes of Englishness: The 'Exceptional' Ideology of Anthony Mario Ludovici," *Journal of Political Ideologies* 4, no. 2 (1999): 191–218.

11. R. W. Connell, "Men's Bodies," in *Which Way Is Up?* (Sydney: George Allen and Unwin, 1983).

12. Sitwell, *Great Morning,* 234.

13. E. M. Forster, *Howards End* (New York: Vintage Books, 1989 [1910]), 24. See also Andrew Thacker, *Moving through Modernity: Space and Geography in Modernism* (Manchester: Manchester University Press, 2003), 62–74; Lionel Trilling, *E. M. Forster: A Study* (London: Hogarth Press, 1969), 109.

14. T. S. Eliot's *The Wasteland* (1922), lines 215–17.

15. Kenneth Grahame, *The Wind in the Willows* (New York: Scribner, 1960 [1908]), 112–13.

16. Evelyn Waugh, *Vile Bodies* (London: Chapman and Hall, 1930), 160.

17. Valentine Cunningham, *British Writers of the Thirties* (Oxford: Oxford University Press, 1988), 161.

18. John C. Moore, *The Walls Are Down* (London: J. M. Dent, 1933), 236. Additional proof of Harry's masculinity is offered by his disparaging reference to "going all pansy."

19. R. Denne Waterhouse, *Week-End Ticket* (London: Arrowsmith, 1934), 101–6.

20. Sigmund Freud, "Three Essays on the Theory of Sexuality" (1905), in *Sigmund Freud on Sexuality* (London: Penguin, 1977), 121

21. Alfred Adler et al., *Individual Psychology and Sexual Difficulties* (London: C. W. Daniel, 1932), 36–37.

22. Michael Balint, *Thrills and Regressions* (London: Hogarth Press, 1959), 23.

23. Wolfgang Sachs, *For Love of the Automobile: Looking Back into the History of our Desires* (Berkeley: University of California Press, 1992), 112–15.

24. Discussing Bram Stoker's *Lady Athlyne* (1908), one critic comments, "In keeping with Stoker's habitual use of mechanical imagery, the unnamable experience of sexual excite-

ment is displaced onto the thrill of speed as Athlyne whisks Joy away on a clandestine ride through the Lake District in his fine red motorcar." David Glover, *Vampires, Mummies, and Liberals: Bram Stoker and the Politics of Popular Fiction* (Durham, NC: Duke University Press, 1996), 129.

25. Aldous Huxley, *Those Barren Leaves: A Novel* (London: Chatto and Windus, 1950 [1925]).

26. Aldous Huxley, *Point Counter Point* (London: Penguin, 1967 [1928]), 346, 347, 371.

27. Patrick Hamilton, *Hangover Square* (London: Penguin, 2001 [1941]), 128. Patrick Hamilton—who was himself run over by a car—envisaged a world in which automobiles become a conquering race. He ends one novel with the vision of "beetles" taking complete control of the countryside. At start of the century, they had been "rare, explosive, laughably crude," but they change and defeat men who now have to satisfy their demands. Patrick Hamilton, *Mr Stimpson and Mr Gorse* (London: Constable, 1953), 353.

28. Clay McShane, *Down the Asphalt Path: The Automobile and the American City* (New York: Columbia University Press, 1994), 139.

29. Forster, *Howards End,* 23.

30. Gilbert Frankau, *Farewell Romance* (New York: E. P. Dutton, 1936), 219.

31. D. H. Lawrence, *The Virgin and the Gypsy* (New York: World Press, 1944 [1930]), 97.

32. D. H. Lawrence, *Women in Love* (New York: Viking, 1960 [1920]), 295.

33. D. H. Lawrence, "In Love," in *The Woman Who Rode Away and Other Stories* (London: Martin Secker, 1928), 198.

34. Sachs, *For Love of the Automobile,* 40.

35. Sitwell, *Great Morning,* 234.

36. Beth Bailey, *From Front Porch to Back Seat: Courtship in Twentieth-Century America* (Baltimore: Johns Hopkins University Press, 1988); Roger N. Casey, *Textual Vehicles: The Automobile in American Literature* (New York: Garland, 1997), 6.

37. See also Virginia Scharff, *Taking the Wheel: Women and the Coming of the Motor Age* (New York: Free Press, 1991), 138–40.

38. Robert S. Lynd and Helen Merrell Lynd, *Middletown: A Study in American Culture* (New York: Harcourt, Brace and World, 1929), 114, 137–38, 258.

39. Bertrand Russell, *Marriage and Morals* (London: George Allen, 1929), 126.

40. Paul R. Deslandes, *Oxbridge Men: British Masculinity and the Undergraduate Experience, 1850–1920* (Bloomington: Indiana University Press, 2005), 110; and see also Sonja Levsen, "Constructing Elite Identities: University Students, Military Masculinity and the Consequences of the Great War in Britain and Germany," *Past and Present* 198 (2008): 162.

41. W. J. McCormack and Patrick Gillan, eds., *The Blackwell Companion to Modern Irish Culture* (Oxford: Blackwell Publishing, 2001), 616.

42. Paul Ferris, *Sex and the British: A Twentieth-Century History* (London: Michael Joseph, 1993), 134.

43. Waugh, *Vile Bodies,* 220–21.

44. Graham Greene, *Brighton Rock* (London: Everyman's Library, 1993 [1938]),164.

45. Graham Greene, "A Drive in the Country" (1937), in *Twenty-One Stories* (London: William Heinemann, 1947).

46. By the 1930s, there were reports of motorcars used for prostitution but only in 1985 did the Kerb Crawling Act specifically target men soliciting from cars. Helen J. Self, *Prostitution, Women and Misuse of the Law: The Fallen Daughters of Eve* (London: Routledge, 2003), 118, 277.

47. F. Tennyson Jesse and H. M. Harwood, *The Pelican* (London: Ernest Benn, 1926), 21.

48. Greene, *Brighton Rock,* 76.

49. Greene, *Brighton Rock,* 122.

50. Sarah S. Jain, "Violent Submission: Gendered Automobility," *Cultural Critique* 61 (Fall 2005): 197; and on male-male desire being channeled through a woman, see Eve Kosofsky Sedgwick, *Between Men: English Literature and Male Homosocial Desire* (New York: Columbia University Press, 1985).

51. Evelyn Waugh, *Brideshead Revisited* (Boston: Little Brown, 1946 [1944]), 23.

52. Angus McLaren, *Sexual Blackmail: A Modern History* (Cambridge, MA: Harvard University Press, 2002), 116–17.

53. *The Times*, March 28, 1932, 7.

54. *The Times,* August 23, 1938, 7; October 5, 1938, 8.

55. Cited in Lord Horder, "The Strain of Modern Civilization," *Nature,* September 26, 1936, 530.

56. Lord Horder was here lifting a line from Ortega y Gasset's *Revolt of the Masses* (1930).

57. Sean O'Connell, *The Car in British Society: Class, Gender and Motoring, 1896–1939* (Manchester: Manchester University Press, 1998); Ross McKibbin, *Classes and Cultures: England, 1918–1951* (Oxford: Oxford University Press, 1998), 58, 109n8.

58. Douglas Goldring, *Odd Man Out* (1935), cited in D. J. Taylor, *Bright Young People: The Rise and Fall of a Generation, 1918–1940* (London: Chatto and Windus, 2007); and see also Philip Ziegler, *Diana Cooper* (London: H. Hamilton, 1981), 160.

59. Martin Wiener, *English Culture and the Decline of the Industrial Spirit, 1850–1980* (Cambridge: Cambridge University Press, 1981), 61.

60. C. E. M. Joad, "The People's Claim," in *Britain and the Beast,* ed. Clough Williams-Ellis (London: Readers' Union, 1938), 73.

61. Forster, *Howards End,* 24.

62. H. G. Wells, *In the Days of the Comet* (London: Macmillan, 1906), 78–79.

63. Cars also figured in music-hall routines. Harry Tate (Ronald Hutchison) had an enormous success with his "Motoring" sketch, in which a chauffeur and his idiotic son fail to get a car started. Harry Tate, *"Motoring": A Farce* (London: Reynolds, 1918).

64. Andrew Horrall, *Popular Culture in London, 1890–1918: The Transformation of Entertainment* (Manchester: Manchester University Press, 2001), 73.

65. Trevor Rowley, *The English Landscape in the Twentieth Century* (London: Hambledon, 2006), 26; Paul Thompson, "Imagination and Passivity in Leisure: Coventry Car Workers and Their Families from the 1920s to the 1970s," in *The Motorcar and Popular Culture in the*

Twentieth Century, ed. David Thoms, Len Holden, and Tim Claydon (Aldershot: Ashgate, 1998), 244–74.

66. Waugh, *Vile Bodies*, 197

67. Greene, *Brighton Rock*, 123.

68. McShane, *Down the Asphalt Path*, 153, 155. On machines and men, see Christopher Forth, *Masculinity in the Modern West: Gender, Civilization and the Body* (London: Palgrave Macmillan, 2008), 177–82.

69. Such relationships could become emotionally charged. Proust fell in love with his chauffeur, Alfred Agostinelli. See William C. Carter, *Proust in Love* (New Haven, CT: Yale University Press, 2006), 72, 113, 128.

70. George Bernard Shaw, *Man and Superman: A Comedy and a Philosophy* (London: Archibald Constable, 1903), 48.

71. Trilling, *E. M. Forster: A Study*, 109.

72. H. G. Wells, *The Wife of Sir Isaac Harman* (London: Macmillan, 1914).

73. Stella Gibbons, *Nightingale Wood* (London: Longmans, Green, 1938), 341. The comic works of P. G. Wodehouse are littered with chauffeurs such as Slingsby, Voules, and Waterbury.

74. Tracy Louis, *Sylvia's Chauffeur* (London: Ward, Lock, 1911); and for similarly transparent plots, see Alice Muriel Williamson and Charles Norris Williamson, *My Friend the Chauffeur* (London: Newnes' Sixpenny Copyright Novels, 1912); Ethel M. Dell, "The Experiment," in *The Safety Curtain, and Other Stories* (London: T. Fisher Unwin, 1917); Scharff, *Taking the Wheel*, 18–22.

75. F. Tennyson Jesse, ed., *Trial of Alma Victoria Rattenbury and George Percy Stoner* (London: W. Hodge, 1950). Terrence Rattigan subsequently based a 1976 play on the case. See Terrence Rattigan, *Cause célèbre* (London: Hamish Hamilton, 1978).

76. Berta Ruck, *Sir or Madame?* (Leipzig: Bernhard Tauchnitz, 1923).

77. On the linkage of lesbians and ambulance drivers, see Tirza True Latimer, *Women Together/Women Apart: Portraits of Lesbian Paris* (New Brunswick, NJ: Rutgers University Press, 2005), 21–22; Scharff, *Taking the Wheel*, 107.

78. Radclyffe Hall, *The Well of Loneliness* (London: Wordsworth Editions, 2006 [1928]), 33, 169, 296.

79. A later French case demonstrates that just as for some men the car replaced the woman, for some women the car replaced the man. Simone de Beauvoir told her ex-lover how she spent the proceeds from *The Second Sex* (1951): "'A woman cannot live without some passion,' she wrote to Nelson Algren. 'As love is forbidden, I decided to give my dirty heart to something not so piggish as a man: and I gave myself a nice beautiful black car.'" Hazel Rowley, *Tête-à-tête: Simone de Beauvoir and Jean-Paul Sartre* (New York: HarperCollins, 2005), 208.

80. Waugh, *Vile Bodies*, 95; and on same-sex attractions and cars, see also Rosamund Lehmann, *Dusty Answer* (London: Chatto & Windus, 1927), 186. In 1920, "Joe" Carstairs began the "X Garage," a London chauffeur service, which only employed female drivers. For the biography of this lesbian sportswoman (and one-time ambulance driver) whose triumphs won her the devotion of hordes of female fans, see Kate Summerscale, *The Queen*

of Whale Cay: The Eccentric Story of "Joe" Carstairs, Fastest Woman on Water (New York: Viking, 1998).

81. Virginia Woolf, *Orlando: A Biography* (London: Hogarth Press, 1928), 268–69, 282.

82. *The Letters of Virginia Woolf,* ed. N. Nicholson and J. Trautmann (London: Hogarth Press, 1977), 3:463, cited in Andrew Thacker, "Traffic, Gender, Modernism," *Sociological Review* 54, no. 1 (2006): 182.

83. Peter Thorold, *The Motoring Age: The Automobile and Britain, 1896–1939* (London: Profile Books, 2003), 125; and see also Georgine Clarsen, *Eat My Dust: Early Women Motorists* (Baltimore: Johns Hopkins University Press, 2008).

84. Virginia Nicholson, *Among the Bohemians: Experiments in Living, 1900–1939* (New York: William Morrow, 2002); *New York Times,* September 15, 1927.

85. It is worth noting that when Dorothy Levitt—in 1903 the first British woman to win a major automobile race—wrote an advice book, *The Woman and Her Car* (1909), her male editor felt it necessary to assure readers in the introduction that Levitt, though she had set speed records, was no Amazon. She was, he insisted, "the most girlish of womanly women. Slight in stature, shy, shrinking, almost timid in everyday life." Julie Wosk, *Women and the Machine: Representations from the Spinning Wheel to the Electronic Age* (Baltimore: Johns Hopkins University Press, 2001), 137.

86. John Buchan, *Greenmantle* (London: Thomas Nelson and Sons, 1956 [1916]), 245.

87. Joshua Zeitz, *Flapper: A Madcap Story of Sex, Style, Celebrity, and the Women Who Made America Modern* (New York: Crown Publishers, 2006); Cynthia Golomb Dettelbach, *In the Driver's Seat: The Automobile in American Literature and Popular Culture* (Westport, CT: Greenwood Press, 1976); Vivien Green-Fryd, *Art and the Crisis of Marriage* (Chicago: University of Chicago Press, 2003), 187–88.

88. Georges Duhamel, *Scènes de la vie future* (Paris: Mercvre de France, 1956 [ca. 1930]), 80; see also *America the Menace: Scenes from the Life of the Future*, trans. Charles Miner Thompson (London: George Allen and Unwin, 1930).

89. Evelyn Waugh, *Black Mischief* (Boston: Little Brown, 1932), 147. Both the cars and the women would be thin and streamlined. "The slippage between descriptions of cars and women's bodies by the thirties reveals the dominance of the discourse of the modern aesthetic." Penny Tinkler and Cheryl Krasnick Warsh, "Feminine Modernity in Interwar Britain and North America: Corsets, Cars, and Cigarettes," *Journal of Women's History* 20, no. 3 (2008): 126.

90. Virginia Woolf, "Professions for Women," in *The Death of the Moth and Other Essays* (London: Hogarth Press, 1942), 151

91. Fred Pape was Woolf's chauffeur and driving instructor, but she gave up trying to learn to drive. *The Diary of Virginia Woolf,* ed. A. Oliver Bell and A. McNeillie. (Harmondsworth: Penguin, 1977), 3:146–47.

92. *The Diary of Virginia Woolf,* 3:151; and see also Andrew Thacker, *Moving through Modernity: Space and Geography in Modernism* (Manchester: Manchester University Press, 2003), 171–83.

93. Harry Joseph Birnstingl, *Lares et Penates; or, The Home of the Future* (London, Kegan, Paul, Trench, Trübner, 1928), 60.

94. Clough Williams-Ellis, ed., *Britain and the Beast* (London: Readers' Union, 1938), 196.

95. Michael Arlen, *The Green Hat: A Romance for a Few People* (London: W. Collins and Sons, 1924), 154, 174.

96. Arlen, *The Green Hat,* 5

97. Arlen, *The Green Hat,* 47

98. In Waugh's *Vile Bodies,* driving eventually leads to a female character's insanity and death.

99. *The Times,* March 1 and 5, 1929.

100. Norah C. James, *Sleeveless Errand* (Paris: Henry Babou and Jack Kahane, 1929), 66.

101. James, *Sleeveless Errand,* 206

102. James, *Sleeveless Errand,* 217. The later American film *Thelma and Louise* (1991, dir. Ridley Scott) similarly ends with a suicidal crash.

103. Elizabeth Bowen, *To the North* (London, J. Cape, 1950 [1932]).

104. John Camden Hotten, *The Slang Dictionary; or, The Vulgar Words, Street Phrases, and "Fast" Expressions of High and Low Society* (London: J. C. Hotten, 1869), 131.

105. Noël Coward presented a comic account of two drunks discussing the issue:

> LEO: How astonished our insides must be—all that brandy hurtling down suddenly! . . . . We ought to know more about our insides, Otto. We ought to know why everything does everything.
>
> OTTO: Machines! That's what we are, really—all of us! I can't help feeling a little discouraged about it now and then. . . . .
>
> LEO: That's because you don't understand; because you're still chained to stale illusions. Science dispels illusions; you ought to be proud to be living in a scientific age.

Noël Coward, *Design for Living,* in *Six Plays* (London: Heinemann, 1934), 86.

106. Nicholas Daly, *Literature, Technology, and Modernity, 1860–2000* (Cambridge: Cambridge University Press, 2004), 117.

107. Sigmund Freud, "Three Essays on the Theory of Sexuality" (1905), in *Sigmund Freud on Sexuality* (London: Penguin, 1977), 122.

108. William J. Robinson, *A Practical Treatise on the Causes, Symptoms, and Treatment of Sexual Impotence* (New York: Critic and Guide, 1923), 153.

109. Victor G. Vecki, *Sexual Impotence.* (Philadelphia and London: W. B. Saunders, 1912), 344.

110. Marie Carmichael Stopes, *Enduring Passion* (London: G. P. Putnam's Sons, 1928), 60.

111. Gladys M. Cox, *Youth, Sex and Life* (London: C. Arthur Pearson, 1935), 201.

112. Claud Mullins, *Marriage, Children and God* (London; George Allen and Unwin, 1933), 184.

113. The explorer-soldier Francis Younghusband, being too timid to describe intercourse clearly, asserted that it was no more than "a simple piston-like operation." See Francis

Younghusband, *Wedding* (1942), cited in Patrick French, *Younghusband: The Last Great Imperial Adventurer* (London: Harper Collins, 1994), 386.

114. Cited in Eustace Chesser, *Love and Marriage* (London: Pan Books, 1957), 73.

115. Charlotte Haldane, *Man's World* (London: Chatto and Windus, 1926), 36; and see also Arnold Bennett, *The Human Machine* (London: New Age Press, 1908).

116. Charles Clinton, *Married Sweethearts: The Rôle of Sex Behavior in Marriage* (New York: MacFadden Book, 1933), 145.

117. Margaret Sanger, *Happiness in Marriage* (New York: Brentano's, 1926), 123; J. F. Tuthill, "Impotence," *Lancet* 265 (January 15, 1955): 126.

118. Isabel Emslie Hutton, *The Hygiene of Marriage* (London: William Heinemann, 1923), 67.

119. Kenneth Walker, "The Art of Love," in *Sex in Social Life,* ed. Sybil Neville-Rolfe (London: George Allen and Unwin, 1949), 345.

120. David Mace, *Sexual Difficulties in Marriage* (London: National Marriage Guidance Council, 1983), vi. The car analogy was also employed by the American eugenicist and marriage expert Paul Popenoe: "As Popenoe, ever the pragmatist was fond of saying, 'if your automobile broke down, you knew where to go for help' but if your marriage collapsed, there was almost nowhere to turn." Alexandra Minna Stern, *Eugenic Nation: Faults and Frontiers of Better Breeding in Modern America* (Berkeley: University of California Press, 2005), 160.

121. Nicholas Daly, *Literature, Technology, and Modernity, 1860–2000* (Cambridge: Cambridge University Press, 2004), 135.

122. Bertrand Russell, *Has Religion Made Useful Contributions to Civilization?* (Girard, KS: Haldeman-Julius Publications, 1929), 22.

123. Julian Huxley, "Searching for the Elixir of Life," *Century Illustrated Monthly Magazine,* February 1922, 626, cited in Susan Merrill Squier, *Babies in Bottles: Twentieth-Century Visions of Reproductive Technology* (New Brunswick, NJ: Rutgers University Press, 1994), 40. On doctors running "repairing shops" like garages, see also Arthur Keith, *The Engines of the Human Body* (London: Williams and Norgate, 1925 [1919]).

124. John Galsworthy, *In Chancery* (New York: Read Books, 2008 [1920]), 7.

125. Walter M. Gallachan, *The Great Unmarried* (London: T. Werner Laurie, 1916), 154–55; and see also Walter M. Gallachan, *The Modern Woman and How to Keep Her* (London: T. Werner Laurie, 1905), 53.

126. Philip Mairet, *Autobiographical and Other Papers,* ed. C. H. Sisson (Manchester: Carcanet, 1981), 203.

127. The working class long used railway analogies in describing coitus interruptus, as in "I'm getting off at Broadgreen" (the station before Edge Hill, Liverpool), but in the 1930s motoring references emerged. A Welshman said of the withdrawal method that it was "like a motorcar he won't go without petrol, will he?" Kate Fisher, *Birth Control, Sex and Marriage in Britain, 1918–1960* (Oxford: Oxford University Press, 2006), 181; and see also Gigi Santow, "Coitus Interruptus and the Control of Natural Fertility," *Population Studies* 49 (1995): 36–38.

128. Marie C. Stopes, *Marriage in My Time* (London: Rich and Cowan, 1935), 233.

129. Enid Charles, *The Twilight of Parenthood: A Biological Study of the Decline of Population Growth* (London: Watts, 1934), 106.

130. Reginald Pound, *A. P. Herbert: A Biography* (London: Michael Joseph, 1976), 150–51

131. Alice Jenkins, *Conscript Parenthood? The Problem of Secret Abortion* (London: George Standring, 1938), 41, 56.

132. Caitriona Beaumont, "Moral Dilemmas and Women's Rights: The Attitude of the Mothers' Union and Catholic Women's League to Divorce, Birth Control and Abortion in England, 1928–1939," *Women's History Review* 16, no. 4 (September 2007): 465; and for a response, see Edith Clara Summerskill, *Wanted—Babies: A Trenchant Examination of a Grave National Problem* (London: Gainsborough Press, 1943).

133. Charles Galton Darwin, *The Next Million Years* (London: Rupert-Hart Davis, 1952), 136.

134. Ferdynand Zweig, *The Worker in an Affluent Society* (London, 1961), 155, cited in Hera Cook, *The Long Sexual Revolution: English Women, Sex, and Contraception, 1800–1975* (Oxford: Oxford University Press, 2004), 324.

135. Foreword written by Brian Stableford for *S. Fowler Wright's Short Stories* (Ludlow: FWB, 1996), 18.

136. J. B. S. Haldane noted that "a materialist is supposed to be a man who has, or does his best to have, large meals, a large mistress, a large bank balance, a large motorcar, and so on." *Science and Life: Essays of a Rationalist* (London: Pemberton Publishing, 1968), 27–28.

137. D. H. Lawrence, *Lady Chatterley's Lover* (New York: Grove Press, 1957 [1928]), 260.

138. David Ellis, *D. H. Lawrence, Dying Game, 1922–1930: The Cambridge Biography of D. H. Lawrence* (Cambridge: Cambridge University Press, 1998), 493.

139. Henry Durant, *The Problem of Leisure* (London: George Routledge and Sons, 1938), 23. For attacks on "The Leisure State," see also Eric Gill, *Last Essays* (London: Jonathan Cape, 1942).

140. Wilfred Wellock, *A Mechanistic or a Human Society?* (London: P.P.U., 1943), 9.

141. George Orwell, "The Rediscovery of Europe" (1942), in *George Orwell: The Collected Essays, Journalism and Letters,* ed. Sonia Orwell and Ian Angus (London: David R. Godine Publisher, 2000), 201.

142. Winifred Holtby, "The Psychology of Revivalism," *The Realist* 1 (April–June 1929): 60.

143. Jonathan Dollimore, "Vehicles of Dissatisfaction," *London Review of Books,* July 24, 2003, 19.

CHAPTER 3

1. For American responses to the play that opened at the Garrick Theatre in New York City in October 1922, see John Corbin, "A Czecho-Slovak Frankenstein," *New York Times,* October 10, 1922, 16.

2. *The Times*, September 12, 1930, 15.

3. *The Times*, August 18, 1932, 14; *Illustrated London News*, August 27, 1932, 301. On an earlier machine, "Eric the Robot" built by Captain W. H. Richards, his "father," which pur-

portedly lifted its arms and answered questions, see *The Times*, September 9, 1929, 9; *New York Times*, September 16, 1928, 26.

4. Joseph Čapek and Karel Čapek, *R.U.R. and The Insect Play*, trans. Paul Selver (London: Oxford University Press, 1961). Selver took liberties with the play, but as his version is the one that the British public saw, it will be the focus of the following discussion. For a new translation, see *Toward the Radical Center: A Karel Čapek Reader*, ed. Peter Kussi (Highland Park, NJ: Catbird Press, 1990).

5. Čapek always capitalizes Robot; for simplicity's sake in what follows it will be spelled in lowercase letters. For photographs of the 1923 cast, see *Illustrated London News*, May 12, 1923.

6. The audience could not fail to be reminded of the recent Russian revolution.

7. On Čapek, see B. R. Bradbrook, *Karel Čapek: In Pursuit of Truth, Tolerance, and Trust* (Eastbourne: Sussex Academic Press, 1998); Ivan Klima, *Karel Capeck: Life and Work*, trans. Norma Comrada (North Haven, CT; Catbird Press, 2002); Jana Horáková and Jozef Kelemán, "The Robot Story: Why Robots Were Born and How They Grew Up," in *The Mechanical Mind in History*, ed. Philip Husbands, Owen Holland, and Michael Wheeler (Cambridge, MA: MIT Press, 2008), 283–306.

8. *The Times*, April 25, 1923, 12; *Daily Mirror*, May 2, 1923, 9.

9. *Daily Mirror*, May 11, 1923, 7; May 14, 1923, 7.

10. *The Times*, September 10, 1928, 4.

11. *Spectator*, June 23, 1923, 1071. Čapek responded in "The Meaning of R.U.R.," *Saturday Review*, July 21, 1923, 79.

12. *The Times*, July 12, 1923, 17.

13. *Daily Mirror*, October 15, 1927, 3.

14. See, for example, *The Times*, December 6, 1928, 16; October 30, 1931, 11.

15. *Daily Mirror*, May 8, 1930, 9.

16. Richard Aldington, *Death of a Hero* (London: Hogarth Press, 1965 [1929]), 228.

17. J. F. C. Fuller, *The Dragon's Teeth: A Study of War and Peace* (London: Constable, 1932), 297.

18. *Daily Mirror*, February 7, 1929, 7.

19. Mabel Broughton Billett, *The Robot Detective* (London: Hutchinson, 1932), 52.

20. *The Times*, August 1, 1929, 15.

21. Jane Goodall, "Transferred Agencies: Performance and the Fear of Automatism," *Theatre Journal* 49, no. 4 (1997): 441–53. The Tiller Girls—a popular English dance troupe—was cited by the Weimar cultural critic Siegfried Kracauer as an example of capitalism producing by assembly line methods new female bodies disciplined to perform geometric dance routines. Siegfried Kracauer, *The Mass Ornament: Weimer Essays*, trans. Thomas Y. Levin (Cambridge, MA: Harvard University Press, 1995), 75–86.

22. *The Times*, March 6, 1930, 12; October 18, 1928, 14. For a novel in which a theater manager experiments with robots, but concludes that is easier to use real actors drilled as subservient slaves, see Robert Graves, *Antigua, Penny, Puce* (Toronto: Macmillan of Canada, 1936), 62–63, 71–72.

23. "Robot Feelings," *The Complete Poems of D. H. Lawrence* (New York: Viking Press, 1971), 3:648, and see also "The Cross," 3:636–37.

24. *The Complete Poems of D. H. Lawrence,* 3:651, 625.

25. D. H. Lawrence, *Lady Chatterley's Lover* (New York: Grove Press, 1959 [1928]), 42.

26. *The Times*, March 2, 1932, 13.

27. Wyndham Lewis, "Shropshire Lads or Robots Again," *New Britain*, January 10, 1934, 226–27, reprinted in *Creatures of Habit and Creatures of Change: Essays on Art, Literature and Society, 1913–1956*, ed. Paul Edwards (Santa Rosa, CA: Black Sparrow Press, 1989), 190–94.

28. T. S. Eliot's "Whispers of Immortality" (1920), cited by Edward J. O'Brien, *The Dance of the Machines: The American Short Story and the Industrial Age* (New York: Macaulay, 1929), 232–33.

29. J. R. R. Tolkien, "On Fairy Stories" (1938), in *The Monsters and the Critics* (London: Allen and Unwin, 1983), 149; and on Tolkien and C. S. Lewis's wariness regarding technology, see Meredith Veldman, *Fantasy, the Bomb and the Greening of Britain* (Cambridge: Cambridge University Press, 1994), 37–114.

30. *The Times,* April 19, 1928, 8. For a cartoon of robot politicians on election platforms giving speeches, being pelted, and breaking down, see *Daily Mirror*, February 6, 1929, 7.

31. Sir A. Daniel Hall, "Knowledge and Power," in *What Science Stands For,* ed. Sir John Boyd Orr et al. (London: George Allen and Unwin, 1937), 105. For the claim that the BBC—in dictating what people should hear—was turning Britain into a nation of robots, see *Daily Mirror*, February 1, 1930, 7.

32. Goodall, "Transferred Agencies: Performance and the Fear of Automatism," 441–53; N. Sharkey and A. Sharkey, "Electro-Mechanical Robots before the Computer," *Journal of Mechanical Engineering Science* 223, no. 1 (2009): 235–41.

33. Percy Bysshe Shelley, *Queen Mab* 3.174–80, in *The Complete Works of Shelley*, ed. Roger Ingpen and Walter E. Peck (London: Ernest Benn, 1965), 1:88.

34. Heinrich Heine, "Concerning the History of Religion and Philosophy in Germany" (1835), in *The Romantic School and Other Essays*, ed. Jost Hermand and Robert C. Holub (New York: Continuum Publishing, 1985), 201.

35. E. T. A. Hoffman, "The Sandman," in *Tales of Hoffman* (New York: Frederick Ungar, 1963), 1–34. That the lead character was based on Lazzaro Spallanzani (1729–1799) who worked on artificial insemination is noted in Paolo Mazzarello, "Sulphur and Holy Water," *Nature* 411 (June 7, 2001): 639.

36. John Cohen, *Human Robots in Myth and Science* (London: George Allen and Unwin, 1967); Allison Muri, *The Enlightenment Cyborg: A History of Communications and Control in the Human Machine, 1660–1830* (Toronto: University of Toronto Press, 2007).

37. Peter Wollen, "Cinema/Americanism/The Robot," in *Modernity and Mass Culture,* ed. James Naremore and Patrick Brantlinger (Bloomington: Indiana University Press, 1991), 47.

38. In *The War of the Worlds* (1898), H. G. Wells describes robot-like fighting devices he calls "Handling Machines," but as they are employed by Martian invaders they are not as frightening as human-made machines that turn on their masters.

39. Harry M. Geduld and Ronald Gottesman, eds., *Robots Robots Robots* (Boston: New York Graphic Society, 1978); Patrick Parrinder, "Robots, Clones and Clockwork Men: The Post-Human Perplex in Early Twentieth-Century Literature and Science," *Interdisciplinary Science Review* 34, no. 1 (2009): 56–67.

40. Čapek and Čapek, *R.U.R. and The Insect Play*, 9.

41. *The Times*, April 25, 1923, 12. On Western culture's fixation on the rebelliousness of created beings from Adam and Eve on, see Lisa Zunshine, *Strange Concepts and the Stories They Make Possible: Cognition, Culture, Narrative* (Baltimore: Johns Hopkins University Press, 2008), 51–54, 85–88.

42. Anson Rabinbach, *The Human Motor: Energy, Fatigue, and the Origins of Modernity* (Berkeley: University of California Press, 1992), 86.

43. John Cunningham Wood, *Alfred Marshall: Critical Assessments* (London: Taylor and Francis, 1993), 293.

44. Frederick W. Taylor, *The Principles of Scientific Management* ((London, Harper and Brothers, 1911). Industrialists had long sought unthinking workers. In the eighteenth century, Adam Ferguson asserted, "Manufactures, accordingly, prosper most, where the mind is least consulted, and where the workshop may, without any great effort of the imagination, be considered as an engine, the parts of which are men." See Frederik Albriton Jonsson, "Enlightened Hands: Managing Dexterity in British Medicine and Manufacture, 1760–1800," in *Body Parts: Critical Explorations in Corporeality*, ed. Christopher E. Forth and Ivan Crozier (Lanham, MD: Lexington Books, 2005), 142–60.

45. Roxanne Panchasi, "Reconstructions: Prosthetics and the Rehabilitation of the Male Body in World War I France," *Differences: A Journal of Feminist Cultural Studies* 7, no. 3 (1995): 109–40; Joanna Bourke, *Dismembering the Male: Men's Bodies, Britain, and the Great War* (Chicago: University of Chicago Press, 1996); Ana Carden-Coyne, *Reconstructing the Body: Classicism, Modernism and the First World War* (Oxford: Oxford University Press, 2009).

46. Bernard Muscio, *Lectures on Industrial Psychology* (London: Routledge, 1920), 182, 244, 247. Born in Australia, Bernard Muscio was educated at Cambridge where in 1919 he became organizing secretary and lecturer for the Industrial Fatigue Research Board. See W. M. O'Neil, "Muscio, Bernard (1887–1926)," *Australian Dictionary of Biography* (1986), 10:650–51.

47. *The Times*, September 17, 1923, 19.

48. Robert W. Dunn, *Labor and Automobiles* (London: Modern Books, 1929), 62.

49. Henry Ford, "Machinery: The New Messiah," in *My Philosophy of Industry* (London: George G. Harrap, 1929), chap. 2.

50. Ronald Campbell Macfie, *Metanthropos; or, The Body of the Future* (London: Kegan Paul, Trench, Trübner 1928), 73.

51. H. Stafford Hatfield, *Automaton; or, The Future of the Mechanical Man* (London: Kegan Paul, Trench, Trübner, 1928), 16.

52. On robots displacing domestic servants, see *Daily Mirror*, March 30, 1928, 9; November 23, 1930, 9; February 7, 1930, 7.

53. Aldous Huxley, *Antic Hay* (New York: Random House, 1932 [1923]), 85.

54. Archibald Montgomery Low, *Our Wonderful World of Tomorrow: A Scientific Forecast of the Men, Women, and the World of the Future* (London: War, Lock, 1934), 250, 254.

55. René Fülop-Miller, *The Mind and Face of Bolshevism*, trans. F. S. Fint and D. F. Tait (London: G. P. Putnam, 1927), 30.

56. Fülop-Miller, *The Mind and Face of Bolshevism*, 211.

57. Kendall E. Bailes, "Alexei Gastev and the Controversy over Taylorism in the Soviet Union, 1920–1924," *Soviet Studies* 29, no. 3 (1977): 373–94.

58. R. Austin Freeman, *Social Decay and Regeneration* (London: Constable, 1921).

59. *The Times*, March 20, 1925, 9.

60. Wilfred Wellock, *A Mechanistic or a Human Society?* (London: P.P.U., 1943).

61. C. E. M. Joad, *Liberty Today* (London: Watts, 1938), 49.

62. *The Times*, May 2, 1933, 16.

63. J. B. Priestly, *English Journey* (London: W. Heinemann, 1949 [1934]), 132.

64. Brian Aldiss and David Wingrove, *Trillion Year Spree: The History of Science Fiction* (New York: Atheneum, 1986), 119.

65. On locating *R.U.R.* in the literary genre of machines gone mad, see Daniel Pick, *War Machine: The Rationalisation of Slaughter in the Modern Age* (New Haven, CT: Yale University Press, 1993), 205–10.

66. The satirist Samuel Butler had earlier played with the notion of machines evolving to the extent of reducing humans to the status of pampered pets. To the assertion that machines could not reproduce, he responded, "What is a reproductive system, if it be not a system for reproduction? And how few of the machines are there which have not been produced systematically by other machines?" Samuel Butler, *Erewhon; or, Over the Range* (1872) (Newark: University of Delaware Press, 1981), 192.

67. James Agate, *The Contemporary Theatre, 1923* (London: Leonard Parsons, 1924), 173.

68. Čapek and Čapek, *R.U.R. and The Insect Play*, 82–83.

69. G. R. Searle, *Eugenics and Politics in Britain, 1900–1914* (Leyden: Noordhoff International, 1976); P. M. Mazumdar, *Eugenics, Human Genetics and Human Failings: The Eugenics Movement, Its Sources and Its Critics in Britain* (London: Routledge, 1992); Christina Cogdell, "Products or Bodies? Streamline Design and Eugenics as Applied Biology," *Design Issues* 19, no. 1 (2003): 42–43.

70. George Bernard Shaw, *Man and Superman: A Comedy and a Philosophy* (London: Archibald Constable, 1903), xxiv.

71. W. R. Inge, "Some Moral Aspects of Eugenics," *Eugenics Review* 1 (1909–10): 26–36.

72. W. C. D. Whetham and Catherine D. Whetham, *The Family and the Nation: A Study in Natural Inheritance and Social Responsibility* (London: Longmans, Green, 1909), 198.

73. Helena M. Swanwick, *The Future of the Women's Movement* (London: G. Bell and Sons, 1913), 200.

74. Huxley, *Antic Hay*, 264.

75. Sydney Fowler Wright, "Automata" (1929), in *S. Fowler Wright's Short Stories* (Ludlow: FWB, 1996), 121–32.

76. O'Brien, *The Dance of the Machines*, 36. The same point was made by a reviewer of

R.U.R.; see John Corbin, "The Revolt against Civilization," *New York Times*, October 15, 1922, 99.

77. Marie C. Stopes, *Marriage in My Time* (London: Rich and Cowan, 1935), 112, 132, 240; and see also A. M. Carr-Saunders, *The Population Problem* (Oxford: Clarendon Press, 1922), 308–16.

78. *Spectator*, May 5, 1923, 755–56.

79. Mrs. M. L. Campbell, "The Automatic Maid-of-All-Work: A Possible Tale of the Near Future," *Canadian Magazine*, July 1893, 394–98.

80. Stephen Leacock, *The Iron Man and the Tin Woman with Other Such Futurities* (London: John Lane The Bodley Head, 1929); and for a review, see *The Times*, November 15, 1929, 20.

81. Čapek and Čapek, *R.U.R. and the Insect Play,* 48–49, 100.

82. Muri, *The Enlightenment Cyborg,* 168. See also M. Norton Wise, "The Gender of Automata in Victorian Britain," in *Genesis Redux: Essays in the History and Philosophy of Artificial Life,* ed. Jessica Riskin (Chicago: University of Chicago Press, 2007), 163–95.

83. Wollen, "Cinema/Americanism/The Robot," 50.

84. Hoffman, "The Sandman," 1–34.

85. Rachilde [Marguerite Eymery], *La Marquise de Sade* (Paris: Gallimard, 1996), 199. In another novel, Rachilde turned the tables—the heroine preserves her dead male lover as a sort of sex toy, a wax mannequin with real teeth and nails, rubber skin, enamel eyes, and a spring set in his mouth that makes him move. Rachilde [Marguerite Eymery], *Monsieur Venus*, trans. Liz Heron (Sawtry: Dedalus, 1992), 144.

86. The novel popularized the term "android." Auguste Villiers de l'Isle Adam, *Tomorrow's Eve*, trans. Robert Martin Adams (Urbana: University of Illinois Press, 1982), 61; and see also Elizabeth K. Menon, "Les Filles d' Ève in Word and Image," in *Writing and Seeing: Essays on Word and Image,* ed. Rui Manuel G. de Carvalho Homem, and Maria de Fátima Lambert (Amsterdam: Rodopi, 2006), 168–69; and Jacques Noiray, *Le Romancier et la machine* (Paris: José Corti, 1981).

87. Thea Von Harbou, *Metropolis* (London: The Readers Library, 1927).

88. Madame B***, avocat, *La Femme endormie* (Melbourne: J. Renold, 1899); and see also René Schwaeblé, *Les Détraqués de Paris, étude de mœurs contemporaines* (Paris: Darangon, 1910), 247–53.

89. Magnus Hirschfeld, *Sexualpathologie* (Bonn: A. Marcus and E. Webers Verlag, 1920), 3:128.

90. J. K. Huysmans, *Against the Grain* (New York: Dover 1969), 162; and see also Rita Felski, *The Gender of Modernity* (Cambridge, MA: Harvard University Press, 1995).

91. But see the postwar film *The Perfect Woman* (1949, dir. Bernard Knowles), which vaguely spoofed the American Lester del Rey's remarkably positive account of two young men (a mechanic and a medical student) setting out to build the perfect domestic and ending up with a loving wife. See Lester del Rey, "Helen O'Loy" (1938), in *The Science Fiction Hall of Fame,* ed. Robert Silverberg (New York: Orb Books, 1998), 1:42–52; Edward James and Farah Mendlesohn, eds., *The Cambridge Companion to Science Fiction* (Cambridge: Cam-

bridge University Press, 2003), 126; Steve Chibnall, "Alien Women: The Politics of Sexual Difference in British SF Pulp Cinema," in *British Science Fiction Cinema,* ed. I. Q. Hunter (London: Routledge, 1999), 58–59.

92. *Daily Mirror*, May 16, 1923, 7.

93. E. V. Odle, *The Clockwork Man* (London: William Heinemann, 1923), 54, 62, 169, 205; and see also *Spectator*, May 19, 1923, 853. The Clockwork Man wears a wig on which is stitched "W. CLARKSON. Wig-Maker to the Seventh International." On Clarkson, see Angus McLaren, "Smoke and Mirrors: Willy Clarkson and the Role of Disguises in Inter-War England," *Journal of Social History* 40, no. 3 (Spring 2007): 597–618.

94. Joseph Storer Clouston, *Button Brains* (London: Herbert Jenkins, 1933). Male robots such as Westinghouse's seven-foot-tall, telephone-controlled "Electro" were apparently sexually exploited. "The Westinghouse robot development stopped with the onset of World War II and never restarted. Electro was restored but then sold onto a Hollywood museum and ended up in seedy movies like 'Sex girls go to college.'" Sharkey and Sharkey, "Electro-Mechanical Robots before the Computer," 237.

95. Čapek and Čapek, *R.U.R. and The Insect Play*, 38.

96. W. H. Bishop, *Education and Heredity; or Eugenics: A Mental, Moral and Social Force* (London: A. and F. Denny, 1909).

97. George Bernard Shaw, *Man and Superman: A Comedy and a Philosophy* (London: Archibald Constable, 1903), 195.

98. Shaw, *Man and Superman*, 125.

99. Bertrand Russell, *Marriage and Morals* (London: George Allen, 1929), 207.

100. R. Ruggles Gates, "A Pedigree Study of Amerindian Crosses in Canada," *Journal of the Royal Anthropological Institute of Great Britain and Ireland* 58 (July 1928): 530; and see also R. Ruggles Gates, "Blood Groups and Physiognomy of British Columbia Coastal Indians," *Journal of the Royal Anthropological Institute* 64 (1934): 40.

101. Sybil Neville-Rolfe, *Why Marry?* (London: Faber and Faber, 1935), 73. See Angelique Richardson, "Rolfe, Sybil Katherine Neville- (1885–1955)," *Oxford Dictionary of National Biography* (Oxford: Oxford University Press, 2004), http://www.oxforddnb.com/view/article/76744, accessed November 4, 2009.

102. Elizabeth Sloan Chesser, *Vitality: A Book on the Health of Women and Children* (London: Methuen, 1935), 61.

103. Nora H. March, *Towards Racial Health* (London: George Routledge and Sons, 1918).

104. H. G. Wells, *A Modern Utopia* (London: Chapman and Hall, 1905). Wells returned to the issue in *Men Like Gods,* in *The Sleeper Awakes and Men Like Gods* (London: Oldham Press, 1923).

105. Elizabeth Sloan Chesser, *Woman, Marriage and Motherhood* (London: Cassell, 1913).

106. Arabella Kenealy, *Feminism and Sex-Extinction* (London: T. Usher Unwin, 1920; Ettie A. Hornibrook [Rout], *Restoration Exercises for Women* (London: William Heineman, 1931); Gladys M. Cox, *Youth, Sex and Life* (London: C. Arthur Pearson, 1935), Ethel Mannin, *Women and the Revolution* (London: Secker and Warburg, 1938).

107. Charlotte Haldane, *Man's World* (London: Chatto and Windus, 1926), 59.

108. *The Stepford Wives* (1975, dir. Bryan Forbes).

109. J. B. S. Haldane, *Daedalus; or, Science and the Future* (London: Kegan Paul, Trench, Trübner, 1924), 40–41.

110. Freeman, *Social Decay and Regeneration*, 303.

111. Čapek and Čapek, *R.U.R. and The Insect Play*, 95.

112. John Macnicol, "The Voluntary Sterilization Campaign in Britain," *Journal of the History of Sexuality* 2 (1992): 422–37. Lancelot Hogben caustically observed: "The eugenic movement has recruited its members from the childless rentiers—twentieth-century Bourbons who have earned nothing and begotten nothing. Its voluptuous insistence on mutilation as the goal of applied genetics has borne fruit in no outstanding discoveries." Lancelot Hogben, *Science for the Citizen* (London: George Allen and Unwin, 1938), 1074.

113. S. A. K. Strahan, *Marriage and Disease* (London: Kegan Paul, Trench, Trübner, 1892).

114. Anthony Ludovici, *Lysistrata; or, Woman's Future and Future Woman* (London: Kegan Paul, Trench, Trübner, 1924); and see also Dan Stone, "The Extremes of Englishness: The 'Exceptional' Ideology of Anthony Mario Ludovici," *Journal of Political Ideologies* 4, no. 2 (1999), 191–218.

115. James Jeans, *More Points of View: A Second Series of Broadcast Addresses* (London: George Allen and Unwin, 1930), 59.

116. Jeans, *More Points of View*, 62. Others disdained the eugenicists' ambitions "to grade great men as one grades prize hogs." See the anthropologist R. H. Lowie, *Are We Civilized? Human Culture in Perspective* (London: George Routledge and Sons, 1929), 286.

117. Elizabeth Sloan Chesser, *Woman, Marriage and Motherhood* (London: Cassell, 1913), 200–212.

118. Neville-Rolfe, *Why Marry?* 139.

119. Greta Jones, *Social Hygiene in Twentieth-Century Britain* (London: Croom Helm, 1986), 106–7.

120. John Desmond Bernal, *The World, Flesh, and the Devil* (London: Kegan Paul, Trench, Trübner, 1929).

121. Bertrand Russell, *The Scientific Outlook* (London: George Allen and Unwin, 1931); and see also Low, *Our Wonderful World of Tomorrow*, 224.

122. Victoria E. M. Bennett and Susan Isaacs, *Health and Education in the Nursery* (London: George Routledge and Sons, 1931), 5–6.

123. Claud Mullins, *Marriage, Children and God* (London: George Allen and Unwin, 1933), 106, 165–66.

124. Aldous Huxley, "A Note on Eugenics" (1927), in *Complete Essays*, ed. Robert S. Baker and James Sexton (Chicago: Ivan R. Dee, 2000), 2:284–85.

125. Aldous Huxley, "What Is Happening to Our Population," in *The Hidden Huxley: Contempt and Compassion for the Masses, 1920–1936*, ed. David Bradshaw (London: Faber and Faber, 1944), 147–58; David Bradshaw, "Huxley's Slump: Planning, Eugenics, and the 'Ultimate Need' of Stability," in *The Art of Literary Biography*, ed. John Batchelor (Oxford: Clarendon Press, 1995), 115–28.

126. Čapek and Čapek, *R.U.R. and the Insect Play*, 23.

127. "The Home for Lost Dogs," *Strand Magazine* 1 (June 1891): 652–53.

128. Edwin Black, *War against the Weak: Eugenics and America's Campaign to Create a Master Race* (New York: Four Walls Eight Windows, 2004), 247.

129. Aldous Huxley, *Crome Yellow* (London: Chatto and Windus, 1949 [1921]), 247.

130. James T. Boulton, ed., *The Letters of D. H. Lawrence* (Cambridge: Cambridge University Press, 1979), 1:81.

131. *The Times*, June 14, 1923, 15.

132. H. G. Wells, *A Modern Utopia* (Lincoln: University of Nebraska Press, 1967 [1905]), 143. And on the elimination of defective children, see also F. William Inman, *Biological Politics: An Aid to Clear Thinking* (London: Simpkin Marshall, 1935), 34.

133. Carveth Read, *Natural and Social Morals* (London: Black, 1909), 159.

134. Dan Stone, *Breeding Superman: Nietzsche, Race and Eugenics in Edwardian and Interwar Britain* (Liverpool: Liverpool University Press. 2002), 127.

135. Stone, *Breeding Superman*, 124, 126.

136. Charles Wickstead Armstrong, *The Survival of the Unfittest* (London: C. W. Daniel, 1927), 75. See also the prediction that in the future degenerates would be apprehended before they actually committed their crimes. "Thus it may come about that we shall not hang the murderer, but shall take to the lethal chamber some who have no murders to answer for." George Godwin, *Cain; or, The Future of Crime* (London: Kegan Paul, Trench, Trübner, 1928), 56.

137. Charles Wickstead Armstrong, *Paradise Found; or, Where the Sex Problem Has Been Solved* (London: John Bale, Sons and Danielson, 1936), 99.

138. Nick Kemp, *Merciful Release: The History of the British Euthanasia Movement* (Manchester: Manchester University Press, 2002), 88; and see also Ian Dowbiggin, "'A Prey on Normal People': C. Killick Millard and the Euthanasia Movement in Great Britain, 1930–55," *Journal of Contemporary History* 36, no. 1 (2001): 59–85.

139. Philip Mairet, *Autobiographical and Other Papers*, ed. C. H. Sisson (Manchester: Carcanet, 1981), 200.

140. Jon Turney, *Frankenstein's Footsteps: Science, Genetics and Popular Culture* (New Haven, CT: Yale University Press, 1998), 88–89. The editorial was a response to E. A. Sharpey-Schäfer's statement that now scientists knew "that body so important for the nutritive and reproductive functions of the cell as the nucleus—which may be said, indeed, to represent the quintessence of cell life—possesses a chemical constitution of no very great complexity, so that we may even hope some day to see the material which composes it prepared synthetically." "An Address on the Nature, Origin, and Maintenance of Life," *British Medical Journal* 2 (1912): 591.

141. Helena M. Swanwick, *The Future of the Women's Movement* (London: G. Bell and Sons, 1913), 146, 148, 150, 152.

142. Sir Humphrey Rolleston, "Discussion on Prophylaxis of Mental Disorder," *British Medical Journal* 2 (1925): 781.

143. Aldous Huxley, *Do What You Will* (London: Chatto and Windus, 1929), 266–67.

144. Philip Gibbs, *The Day after To-Morrow: What Is Going to Happen to the World* (London: Hutchinson, 1928), 84.

145. Stanley L. Baker, "Love according to Legislation," *Daily Mirror*, March 10, 1928, 4.

146. John Malcolm, "Shall We Get a Race of Human Machines?" *Daily Mirror*, June 22, 1928, 9. On cautions against the eugenicists' search for a perfect race, see also Mary G. Cardwell, *Some Aspects of Child Hygiene* (London: Pitman and Son, 1935), 14.

147. *Daily Mirror*, August 22, 1932, 7.

CHAPTER 4

1. Aldous Huxley, *Brave New World* (London: Chatto and Windus, 1932), 12, 13, 38, 72, 156, 239–40, 255.

2. F. A. E. Crew review of Paul Kammerer, *Rejuvenation and the Prolongation of Human Efficiency: Experiences with the Steinach Operation on Man and Animals* (London: Methuen, 1924), in *Nature* 114 (October 11, 1924): 530–31. Crew made his eugenic beliefs clear: "Men are not born free and equal but unequal and bound." F. A. E. Crew, *Organic Inheritance in Man* (London: Seeley Service, 1925), 4.

3. *The Times*, July 20, 1922, 5.

4. Noël Coward, *Design for Living* (1932), in *Six Plays* (London: Heinemann, 1934), 5, 6.

5. Susan Merrill Squier, *Babies in Bottles: Twentieth-Century Visions of Reproductive Technology* (New Brunswick, NJ: Rutgers University Press, 1994), 37–38.

6. Stanislas Higier, *Les Fonctions sexuelles mâles et leurs troubles* (Paris: G. Doin, 1932); Diane Long Hall, "Biology, Sex Hormones and Sexism in the 1920's," *Philosophical Forum* 5 (1973–74): 81–96; and for an overview, see Chandak Sengoopta, *The Most Secret Quintessence of Life: Sex, Glands, and Hormones, 1850–1950* (Chicago: University of Chicago Press, 2006).

7. Dr. Brown-Séquard, "The Effects Produced on Man by Subcutaneous Injections of a Liquid Obtained from the Testicles of Animals" *Lancet* 134 (July 20, 1889): 105–6.

8. Arthur T. Davies, "Organo-Therapy," *Lancet* 159 (April 19, 1902): 1089–96; Merriley Borell, "Brown-Séquard's Organotherapy and Its Appearance in America at the End of the Nineteenth Century," *Bulletin of the History of Medicine* 50 (1976): 309–20; Chandak Sengoopta, "Transforming the Testicle; Science, Medicine and Masculinity, 1800–1950," *Medicina nei Secoli Arte e Scienza* 13 (2001): 637–55.

9. Victor Cornelius Medvei, *The History of Clinical Endocrinology: A Comprehensive Account of Endocrinology from Earliest Times to the Present Day* (New York: Parthenon Publishing, 1993), 161.

10. See, for example, Edward Schafer, "Physiology of Internal Secretions," *Lancet* 146 (August 10, 1895): 321–24.

11. *The Times*, June 28, 1923, 11; and see also Diana Long Hall, "The Critic and the Advocate: Contrasting British Views on the State of Endocrinology in the Early 1920s," *Journal of the History of Biology* 9 (1976): 269–85.

12. Serge Voronoff and George Alexandrescu, *Testicular Grafting from Ape to Man*, trans. Theodore C. Merrill (London: Bretano's, 1928), 27, 28; and see also David Boyd, *Mortal Coil: A Short History of Living Longer* (New Haven, CT: Yale University Press, 2008), 175–80.

13. *The Times*, October 9, 1922, 11; and see also October 24, 1923, 13.

14. *Lancet* 197 (May 14, 1921): 1041–42.

15. *The Times*, January 5, 1924, 8; *The Argus* (Melbourne), March 1, 1924, 7.

16. Hamilton Bailey, "Testicular Grafting," *Lancet* 209 (February 5, 1927): 284; *The Times*, November 3, 1926, 7.

17. *Daily Mirror*, August 4, 1931, 6.

18. Voronoff and Alexandrescu, *Testicular Grafting from Ape to Man*, 27, 28; David Hamilton, *The Monkey Gland Affair* (London: Chatto and Windus, 1986); Serge Voronoff, *The Conquest of Life* (London: Williams and Norgate, 1933), 45; *New York Times*, October 3, 1921, 28; and see also Max Hodann, *History of Modern Morals*, trans. Stella Browne (London: Heinemann, 1937), 29; Kammerer, *Rejuvenation and the Prolongation of Human Efficiency*, 63.

19. *The Times*, May 24, 1928, 14; *Daily Mirror*, May 25, 1928, 5.

20. Eugen Steinach and Josef Loebel, *Sex and Life: Forty Years of Biological and Medical Experiments* (London: Faber and Faber, 1940) 21; Carol Sue Carter, *Hormones and Sexual Behavior* (Stroudsburg, PA: Dowden, Hutchinson and Ross Inc., 1974), 16–19.

21. Chandak Sengoopta, "'Dr. Steinach Is Coming to Make Old Young': Sex Glands, Vasectomy and the Quest for Rejuvenation in the Roaring Twenties," *Endeavour* 27, no. 3 (2003): 122–26; Boyd, *Mortal Coil*, 180–87.

22. Steinach and Loebel, *Sex and Life*, 25.

23. The American surgeon Max Thorek found gland transplants could result in a temporary improvement, better described as "reactivation" than rejuvenation. Max Thorek, *The Human Testis* (Philadelphia: J. B. Lippincott, 1924), 285; Max Thorek, *A Surgeon's World, an Autobiography* (Philadelphia: Lippincott, 1943), 190.

24. Francis H. A. Marshall, *The Physiology of Reproduction* (London: Longmans, Green, 1922), 331.

25. *Lancet* 197 (February 26, 1921): 454; *Lancet* 201 (February 24, 1923): 393; R. Mansell-Jones, "The Steinach Operation," *Lancet* 201 (June 16, 1923): 1239; *Lancet* 204 (September 20, 1924): 616.

26. Aldous Huxley, *Antic Hay* (New York: Random House, 1932 [1923]), 348.

27. *Daily Mirror*, December 5, 1929, 2.

28. *The Times*, January 21, 1932, 11.

29. *The Times*, May 14, 1921, 7; and for a photograph of Wilson, see *Daily Mirror*, April 12, 1921, 8.

30. Norman Haire, "How I Run My Birth Control Clinic," in *Sexual Reform Congress, Copenhagen 1–5: VII: 1928, Proceedings of the Second Congress* (Copenhagen: Levin and Munksgaard, 1929), 182–91; and see also Ivan Crozier, "Becoming a Sexologist: Norman Haire, the 1929 London World League for Sexual Reform Congress, and Organizing Medical Knowledge about Sex in Interwar England," *History of Science* 29 (2001): 299–329.

31. British Society for the Study of Sex Psychology, no. 11. E. Paul, *Steinach's Rejuvenation Experiments* and N. Haire, *Recent Developments of Steinach's Work* (London: Athenaeum Press, 1923); and see also Johannes Rutgers, *The Sexual Life*, trans. Norman Haire (Dresden: Verlag R. A. Giescke, 1923), 30.

32. Norman Haire, *Rejuvenation: The Work of Steinach, Voronoff and Others* (London: George Allen and Unwin, 1924), 125; and see also *Eugenics Review* 17 (1925–26): 110–11.

33. *The Times*, September 14, 1929, 12.

34. Peter Schmidt, *The Conquest of Old Age,* trans. Eden and Cedar Paul (London: Routledge, 1931), 117, 119.

35. Kenneth M. Walker, "The Internal Secretion of the Testis," *Lancet* 203 (January 5, 1924): 6–21. For another well-known surgeon who carried out a similar number of ligations, see Hamilton Bailey, "Testicular Grafting," *Lancet* 209 (February 5, 1927): 284.

36. Kenneth M. Walker and J. A. Lumsden, "Steinach's Rejuvenation Operation," *Lancet* 203 (February 2, 1924): 223–26.

37. Kenneth Walker, *Male Disorders of Sex* (London: Jonathan Cape, 1930), 77; and see also Kenneth Walker, *Commentary on Age* (London: Jonathan Cape, 1952), 50–62.

38. Ernest Jones, *The Life and Work of Sigmund Freud* (New York: Basic Books, 1957), 3:104; Diana Wyndham, "Versemaking and Lovemaking: W. B. Yeats' 'Strange Second Puberty': Norman Haire and the Steinach Rejuvenation Operation," *Journal of the History of the Behavioral Sciences* 39 (2003): 25–50; Tim Armstrong, *Modernism, Technology and the Body: A Cultural Study* (Cambridge: Cambridge University Press, 1998), 155.

39. Ethel Mannin, *Young in the Twenties: A Chapter of Biography* (London: Hutchinson, 1971), 67.

40. Alison Li, "Wondrous Transformations: Endocrinology after Insulin," in *Essays in Honour of Michael Bliss: Figuring the Social,* ed. E. A. Heaman, Alison Li, and Shelley McKellar (Toronto: University of Toronto Press, 2008), 351–77.

41. George Ryley Scott, *The Quest for Youth: A Study of All Available Methods of Rejuvenation and of Retaining Physical and Mental Vigor in Old Age* (London: Torchstream, 1953), 9.

42. For an earlier French song satirizing such investigations, see "Le Rénovateur Brown-Séquard," *La Plume* 40 (December 15, 1889): 231–32.

43. Bertram Gayton, *The Gland Stealers* (London: Herbert Jenkins, 1922). The joke is that it is the excitement of the adventure—not the glands—that reinvigorates the old man.

44. *The Times*, February 17, 1926, 12.

45. *Evergreen* (1934, dir. Victor Saville).

46. Norman Haire, *Hymen; or, The Future of Marriage* (London: Kegan Paul, Trench, Trübner, 1927), 88–89.

47. *The Times*, June 8, 1928, 18.

48. See Andrés Horacio Reggiani, *God's Eugenicist: Alexis Carrel and the Sociobiology of Decline* (New York: Berghahn Books, 2006); *The Times*, July 18, 1923, 14.

49. Susan E. Lederer, "Animal Parts/Human Bodies: Organic Transplantation in Early Twentieth-Century America," in *Animal/Human Boundary: Historical Perspectives,* ed. Angela N. H. Creager and William Chester Jordan (Rochester, NY: University of Rochester Press, 2002), 305–29.

50. Kirill Rossiianov, "Beyond Species: Il'ya Ivanov and His Experiments on Cross-Breeding Humans with Anthropod Apes," *Science in Context* 15 (2002): 277–316.

51. George Ryley Scott, *Three Hundred Sex, Marriage and Birth Control Questions Answered* (London: T. Werner Laurie, 1941), 49.

52. Sir Arthur Conan Doyle, *The Penguin Complete Sherlock Holmes* (London: Penguin, 1981), 1082–83; and see Joseph A. Kester, *Sherlock's Men: Masculinity, Conan Doyle, and Cultural History* (Aldershot: Ashgate, 1997), 59.

53. Lord Dunsany [Edward Plunkett], *Lord Adrian, a Play in Three Acts* (Watham Saint Lawrence, Berkshire: Golden Cockerel Press, 1933), 60.

54. *The Times*, March 13, 1937, 10; April 17, 1937, 12.

55. Noël Coward, *Plays* (London: Methuen, 1979), 2:50–51.

56. Dr. L. C. Dundas Irvine, "Rejuvenation: A Reply to Dr. Voronoff," *English Review* 40 (1925): 167–71; Benjamin Harrow, *Glands in Health and Disease* (London: George Routledge and Sons, 1922), 112. For reports from Vienna indicating a decline in the belief in the Steinach operation, see *Lancet* 205 (March 28, 1925): 683.

57. "Voronoff's Method of Rejuvenation," *Lancet* 209 (February 26, 1927): 446–47.

58. *The Times*, October 19, 1923, 8.

59. Bertold Paul Wiesner, *Sex* (London: Thornton Butterworth, 1936), 66, 67; Jean Paul Pratt, "Sex Functions in Man," in *Sex and Internal Secretions: A Survey of Recent Research*, ed. Edgar Allen (London: Baillière, Tindall and Cox, 1939), 1276.

60. Charlotte Haldane, *Nature*, April 23, 1932, 597–98.

61. Gerhard J. Newerla, "The History and Isolation of the Male Hormone," *New England Journal of Medicine* 228 (1943): 39–47; Anne Fausto-Sterling, *Sexing the Body: Gender Politics and the Construction of Sexuality* (New York: Basic Books, 2000), 181–84.

62. Benjamin Harrow and Carl P. Sherwin, *The Chemistry of the Hormones* (London: Baillière, Tindall and Cox, 1934), 152. But for an American enthusiast who claimed treated patients "grew more strong-thinking and less sissified," see Paul de Kruif, *The Male Hormone* (New York: Harcourt, Brace, 1945), 177.

63. British Medical Association, *More Secret Remedies: What They Cost and What They Contain* (London: BMA, 1912); Wellcome Library, CMCS File no. SA/BMA/box 119/ C. 426 Patent Medicines.

64. *Lancet* 208 (December 4, 1926): 1188–89.

65. *The Times,* May 28, 1932, 7. The doctor in question offered in addition the rejuvenation therapy of N. E. Ischlondsky of Paris. See Dr. N. E. Ischlondsky, *Sécrétion interne et régénérescence* (Paris: Doin, 1933).

66. G. W. Carnick Co., *Dependable Gland Products Price List* (London, 1923); Henry R. Harrower, *A Manual of Pluriglandular Therapy* (London: Endocrines Limited, 1924), 239; see also Henry R. Harrower, *Practical Organotherapy: The Internal Secretions in General Practice* (Glendale, CA: Harrower Laboratory, 1922); Middlesex Laboratory of Glandular Research, *The Treatment of Impotence* (London: Middlesex Laboratory of Glandular Research, circa 1930).

67. British Glandular Products, *The Essence of Life* (London, n.d.), 6. For other testicular compounds on offer including Testifortan, Testiglandol, Testogan, Testanon, Spermin Poehl, Dynotabs, and Testacoids, see Theodore van de Velde, *Fertility and Sterility in Marriage: Their Voluntary Promotion and Limitation,* trans. F. W. Stella Browne (London: W. Heinemann, 1931), 255.

68. Dr. H. Schmeidler, *Rejuvenation of Man and Woman* (London: Viriline-Gesellschaft, 1929).

69. Gerald Heard, *Pain, Sex and Time* (London: Harper and Brothers, 1939), 264; Gerald Heard, *The Source of Civilization* (London: Jonathan Cape, 1935), 113.

70. C. E. M. Joad, *Diogenes; or, The Future of Leisure* (London: Kegan Paul, Trench, Trübner, 1928), 90; and on sex determination being the ultimate form of the "mechanization" of sex, see Peter Ritchie-Calder, *Birth of the Future* (London: A. Barker, 1934), 70.

71. Bertrand Russell, *Icarus; or, The Future of Science* (London: Kegan Paul, Trench, Trübner, 1924), 53, 54, 55.

72. *The Times*, February 23, 1922, 11.

73. On reports that pupils in a Chicago subnormal class were to be fed a sheep gland diet with the goal of putting "new brains into the heads of subnormal children," see *New York Times*, October 19, 1921, 21; December 8, 1921, 21. On prisoners and gland transplants, see *New York Times*, February 5, 1922, sec. iii, 9.

74. On the United States, see Julia E. Rechter, "'The Glands of Destiny': A History of Popular, Medical and Scientific Views of the Sex Hormones in 1920s America" (PhD diss., History Department, University of California, Berkeley, 1997), 25.

75. J. B. S. Haldane, *Daedalus; or, Science and the Future* (London: Kegan Paul, Trench, Trübner, 1924), 71.

76. George Godwin, *Cain; or, The Future of Crime* (London: Kegan Paul, Trench, Trübner, 1928).

77. Russell, *Icarus; or, The Future of Science*.

78. Philip Gibbs, *The Day after To-Morrow: What Is Going to Happen to the World* (London: Hutchinson, 1928).

79. Dorothy L. Sayers, *The Unpleasantness at the Bellona Club* (London: Harper and Brothers, 1956 [1928]), 215.

80. *The Times*, January 7, 1938, 8.

81. Edred M. Corner, *Male Diseases in General Practice* (London: Hodder and Stoughton, 1910), 228.

82. Sybil Neville-Rolfe, *Why Marry?* (London: Faber and Faber, 1935), 73.

83. *Eugenical News* 2 (1917): 35; William Engelbach, *Endocrine Medicine* (Baltimore: Charles C. Thomas, 1932), 455–59; and see also Oscar Riddle, "Endocrinology and Eugenics," *Journal of Heredity* 24 (1933): 179–80.

84. *The Times*, March 19, 1924, 11.

85. Engelbach, *Endocrine Medicine*, 442.

86. Louis Berman, *The Glands Regulating Personality* (New York: J. J. Little and Ives, 1921), 288–89.

87. "Endocrine Imbalance in Hybrids," *Eugenical News* 11 (1926): 173.

88. Arthur Keith, *A New Theory of Human Evolution* (New York: Philosophical Library, 1949), 149.

89. Ivo Geikie Cobb, *The Glands of Destiny: A Study of the Personality* (London: W. Heinemann, 1936), 145–65, 253–57.

90. Jon Alfred Mjøen, "Race Crossing and Glands," *Eugenics Review* 33 (1931): 39; and see also Jon Alfred Mjøen, "Biological Consequences of Race Crossing," *Journal of Heredity* 17 (1926): 175–82.

91. *British Medical Journal* 2 (July 21, 1923): 121.

92. Norman Haire, "Rejuvenation," *The Realist* 1 (1929): 125.

93. H. G. Wells, *The Food of the Gods and How It Came to Earth* (New York: Charles Scribner's Sons, 1904), 114.

94. Gibbs, *The Day after To-Morrow*, 59.

95. Charles Percy Snow, *New Lives for Old* (London: Gollancz, 1933); and see David Shusterman, *C. P. Snow* (Boston: Twayne, 1975), 45–47.

96. Charles Galton Darwin, *The Next Million Years* (London: Rupert-Hart Davis, 1952), 183.

97. Fausto-Sterling, *Sexing the Body.*

98. Adele Clarke, *Disciplining Reproduction: Modernity, American Life Sciences, and "the Problems of Sex"* (Berkeley: University of California Press, 1998).

99. Alexander Lipschütz, *The Internal Secretions of the Sex Glands: The Problem of the "Puberty Gland"* (Cambridge: Heffer, 1924), 483–90.

100. Arthur Cooper, *The Sexual Disabilities of Man and Their Treatment* (London: H. K. Lewis, 1908); Lawrence W. Harrison, *The Diagnosis and Treatment of Venereal Diseases in General Practice* (London: Hodder and Stoughton, 1918), 25; William Wynn Westcott, "Sudden and Unexpected Deaths," *British Medical Journal* 1 (February 29, 1908): 491; Gerald Dalton, *A Practical Manual on Venereal and Generative Diseases* (London: Henry Kimpton, 1912), 89; Arthur Cooper, *The Sexual Disabilities of Man and Their Treatment* (London: H. K. Lewis, 1908), 30, 80.

101. *Lancet* 192 (September 14, 1918): 363–64.

102. Kenneth Walker and Eric B. Strauss, *Sexual Disorders in the Male* (London: Hamish Hamilton, 1944), 102.

103. Osbert Sitwell, *Great Morning* (London: Macmillan, 1948), 230.

104. Jean Frumusan, *Rejuvenation* (London: John Bale, Sons and Danielsson, 1923).

105. Nicola Humble, *The Feminine Middlebrow Novel, 1920s to 1950s: Class, Domesticity, and Bohemianism* (Oxford: Oxford University Press, 2001), 204.

106. John Gloag, *Winter's Youth* (London: George Allen and Unwin, 1934), 43–44, 161, 167, 179–81, 195, 246.

107. Lord Birkenhead [Frederick Edwin Smith], *The World in 2030 A.D.* (London: Hodder and Stoughton, 1930), 12.

108. Huxley, *Brave New World*, 66–67.

109. Aldous Huxley, *After Many a Summer* (London: Chatto and Windus, 1939).

110. Chandak Sengoopta, "The Modern Ovary: Constructions, Meanings, Ideas," *History of Science* 28 (2000): 463; Susan Squier, "Incubabies and Rejuvenates: The Traffic between Technologies of Reproduction and Age-Extension," in *Figuring Age: Women, Bodies, Generations*, ed. Kathleen Woodward (Bloomington: Indiana University Press, 1999), 89; Laura Davidow Hirshbein, "The Glandular Solution: Sex, Masculinity, and Aging in the 1920s," *Journal of the History of Sexuality* 9 (2000): 277–304.

111. Haldane, *Daedalus; or, Science and the Future*, 74.

112. *The Times*, May 24, 1928, 14.

113. Sengoopta, "The Modern Ovary," 425–88.

114. Marie Corelli, *The Young Diana: An Experiment of the Future* (London: Hutchinson, 1918).

115. See Harry Benjamin, "The Reactivation of Women," in *Sexual Reform Congress: Proceedings of the Third Congress,* ed. Norman Haire (London: Paul, Trench, Trübner, 1930), 564–73.

116. Gertrude Atherton, *Adventures of a Novelist* (London: Jonathan Cape, 1932), 540.

117. Gertrude Atherton, *Black Oxen* (New York: A. L. Burt, 1923), 172–76. See also Julie Prebel, "Enginering Womanhood: The Politics of Rejuvenation in Gertrude Atherton's Black Oxen," *American Literature* 76, no. 2 (2004): 307–37.

118. David H. Keller, "The Feminine Metamorphosis" *Science Wonder Stories* 1 (1929): 246–60.

119. Vera Brittain, *Halcyon; or, The Future of Monogamy* (London: Kegan Paul, Trench, Trübner, 1929), 73.

120. Mary Scharlieb, *Change of Life: Its Difficulties and Dangers* (London: Scientific Press, 1919), 34–35.

121. Scharlieb, *Change of Life,* 35; and see also John H. Hannan, *The Flushings of the Menopause* (London: Baillière, Tindall and Cox, 1927), 34–36.

122. M. E. Mitchell, *Yet in My Flesh* (London: J. M. Dent, 1933), 27, 46, 287.

123. Squier, "Incubabies and Rejuvenates," 89.

124. George Ryley Scott, *The Quest for Youth: A Study of All Available Methods of Rejuvenation and of Retaining Physical and Mental Vigor in Old Age* (London: Torchstream, 1953), 79.

125. Steinach and Loebel, *Sex and Life,* 250.

126. George Ryley Scott, *Three Hundred Sex, Marriage and Birth Control Questions Answered* (London: T. Werner Laurie, 1941), 116–17.

127. Haire, *Rejuvenation,* 32.

128. See the *Eugenics Review* 18 (1926–27): 351 for a review of Max Schlapp, *Medical and Eugenic Aspects of Birth Control* (New York: ABC, 1926) in which it was argued women's glandular upsets resulted in the production of cretins.

129. On the American experience, see Judith A. Houck, *Hot and Bothered: Women, Medicine, and Menopause in Modern America* (Cambridge, MA: Harvard University Press, 2006).

130. *The Times,* October 25, 1928, 5.

131. Arnold Lorand, *Life Shortening Habits and Rejuvenation* (Montreal: Canadian Medical Book, 1922), 198, 215.

132. *The Times,* June 10, 1936, 19.

133. A. T. Cowie, *Pregnancy Diagnosis Tests: A Review* (Edinburgh: Commonwealth Agricultural Bureaux, 1948); B. M. Hobson, "Pregnancy Diagnosis," *Journal of Reproductive Fertility* 12 (1966): 33–35; Michael J. O'Dowd and Elliot Elias Philipp, *The History of Obstetrics and Gynaecology* (New York: Informa Healthcare, 2000), 85.

134. F. B. Hutt, "Genetics of the University of Edinburgh: The Work of the Department of Animal Genetics," *Journal of Animal Science* (1931): 104; B. P. Wiesner, "Pregnancy Diagnosis Station: Report on Third Year's Working," *British Medical Journal* 2 (October 22, 1932): 759; F. A. E. Crew, "Biological Pregnancy Diagnosis Tests," *British Medical Journal* 1 (April 15 1939): 766–70.

135. Ann Oakley, *The Captured Womb: A History of the Medical Care of Pregnant Women* (Oxford: Blackwell, 1984), 98.

136. Vern Bullough, *Science in the Bedroom: A History of Sex Research* (New York: Basic Books, 1994), 126–32.

137. Nelly Oudshorne, *Beyond the Natural Body: An Archeology of Hormones* (London: Routledge, 1994), 108.

138. Sheila M. Rothman and David J. Rothman, *The Pursuit of Perfection: The Promise and the Perils of Medical Enhancement* (New York: Pantheon, 2003), 41; Elizabeth Siegel Watkins, *The Estrogen Elixir: A History of Hormone Replacement Therapy in America* (Baltimore: Johns Hopkins University Press, 2007).

139. Donald J. Childs, *Modernism and Eugenics: Woolf, Eliot, Yeats, and the Culture of Degeneration* (Cambridge: Cambridge University Press, 2001), 146; Carole Seymour-Jones, *Painted Shadow: A Life of Vivienne Eliot* (London: Constable, 2001), 322, 324.

140. When a science journalist visited his Edinburgh lab in 1934, Wiesner presented him with a pituitary extract hormone—described as "mother-love in a test-tube." Ritchie-Calder, *Birth of the Future*, 65; see also Merriley Borell, "Biologists and the Promotion of Birth Control Research, 1918–1938," *Journal of the History of Biology* 20, no. 1 (1987): 51–87.

141. Cited in *Journal of Social Hygiene* 13 (1927): 142.

142. Ritchie-Calder, *Birth of the Future*, 67.

143. "The Ovarian Hormone," *Nature* 121 (March 31, 1928): 498–99; B. P. Wiesner, "The Hormones and Their Control of the Reproductive System," *Eugenics Review* 22 (1930): 19–26; Merrily Borell, "Organo-Therapy and the Emergence of Reproductive Endocrinology," *Journal of the History of Biology* 18 (1985): 1–30; and on Wiesner's involvement in the production of Emmenin—the first replaceable estrogen, see Alison Li, *J. B. Collip and the Development of Medical Research in Canada* (Montreal: McGill-Queen's University Press, 2003), 67–75.

144. Richard A. Soloway, "The 'Perfect Contraceptive': Eugenics and Birth Control Research in Britain and America in the Interwar Years," *Journal of Contemporary History* 30 (1995): 637–43. Another member of the Eugenics Society who did research on pregnancy testing was the refugee German socialist physician Rudolph Elkan. See Dr. Rudolf Edward Elkan, "The Xenopus Pregnancy Test," *British Medical Journal* 2 (December 17, 1938): 1253–56; Robert Proctor, *Racial Hygiene: Medicine under the Nazis* (Cambridge, MA: Harvard University Press, 1988), 271–72; Atina Grossmann, *Reforming Sex: The German Movement for Birth Control and Abortion Reform, 1920–1950* (Oxford: Oxford University Press, 1997), 24, 30, 134.

145. A doctor attempts to give animals human characteristics via skin grafts, tissue transplants, and blood transfusions in H. G. Wells, *The Island of Dr. Moreau* (London: William Heinemann, 1896).

146. *The Cheetah Girl* bears several resemblances to Hans Heinz Ewers, *Die Alraune: The Story of a Living Creature* (1911), which begins with references to artificial insemination and climaxes with an androgynous girl being born as a result of a professor implanting in a prostitute the seed ejaculated by an executed criminal. To quote Ewers, both works conjured up a "brothel-like atmosphere which cloaked itself in the rarefied aura of scientific discourse." See Hanns Heinz Ewers, *Die Alraune: The Story of a Living Creature* (1911), in

The Dedalus Book of German Decadence: Voices of the Abyss, ed. Raymond Furness (Sawtry, Cambs: Dedalus, 1994), 208.

147. Christopher Blayre [Edward Heron-Allen], *The Cheetah Girl* (London: Printed as a manuscript for private circulation, 1923). On Edward Heron-Allen (1861–1943), friend of Oscar Wilde and expert in the construction of violins, the translation of Persians texts, and paleontology, see R. A. Gregory, "Edward Heron-Allen, 1861–1943," *Obituary Notices of Fellows of the Royal Society* 4, no. 12 (November 1943): 447–54.

148. Niels Hoyer, ed., *Man into Woman: An Authentic Record of a Change of Sex* (London: Jarrolds, 1933); Tim Armstrong, *Modernism, Technology and the Body: A Cultural Study* (Cambridge: Cambridge University Press, 1998), 159–86.

149. *Daily Herald*, May 29, 1936; *Time Magazine*, August 24, 1936, 33; Donald Furthman Wickets [G. S. Viereck?], "Can Sex in Humans Be Changed?" *Physical Culture* 77 (January 1937): 16–18. See also L. R. Broster, Clifford Allen, H. W. C. Vines, Jocelyn Patterson, Alan W. Greenwood, G. F. Marrian, and G. C. Butler, *The Adrenal Cortex and Intersexuality* (London: Chapman and Hall, 1938).

150. On press coverage of Broster's subsequent cases, see *Daily Mirror*, May 5, 1938, 2.

151. Steinach and Loebel, *Sex and Life*, 116–18; Gregorio Marañón, *The Evolution of Sex and Intersexual Conditions,* trans. Warren B. Welles (London: George Allen and Unwin, 1932), 168–69.

152. Berman, *The Glands Regulating Personality*, 222, 226, 248–52.

153. Chris Evans Morris, *Modern Rejuvenation Methods* (New York: Scientific Medical Publishing, 1926), 72; and see also Herman H. Rubin, *Glands, Sex and Personality* (New York: Wilfred Funk, 1952), 73.

154. Robert T. Frank, *The Female Sex Hormone* (London: Baillière, Tindall and Cox, 1929), 115.

155. On Russian scientists' acceptance of the hormonal explanation of homosexuality, see Dan Healey, *Homosexual Desire in Revolutionary Russia: The Regulation of Sexual and Gender Dissent* (Chicago: University of Chicago Press, 2001), 134–36; Frances L. Bernstein, "'The Dictatorship of Sex': Science, Glands, and the Medical Construction of Gender Difference in Revolutionary Russia," in *Russian Modernity: Politics, Knowledge, Practices,* ed. David L. Hoffman and Yanni Kotsonis (New York: Routledge, 2000), 138–60.

156. Julian Huxley, "Sex Biology and Sex Psychology," in *Essays of a Biologist* (London: Chatto and Windus, 1923), 146–47.

157. Dr. H. Schmeidler, *Rejuvenation of Man and Woman* (London: Viriline, n.d.), 30.

158. Neville-Rolfe, *Why Marry*? 104.

159. Magnus Hirschfeld, *Sexual Pathology: A Study of Derangements of the Sexual Instinct,* trans. Jerome Gibbs (New York: Emerson Books, 1947), 310–13; and see Chandak Sengoopta, "Glandular Politics: Experimental Biology, Clinical Medicine, and Homosexual Emancipation in Fin-de-Siècle Central Europe," *Isis* 89 (1998): 448, 456; *Journal of the American Medical Association* 75 (1920): 755; Haire, *Rejuvenation*, 139–46.

160. Haire, *Rejuvenation*, 139–43; and see also Norman Haire, "Rejuvenation," *The Realist* 1 (1929): 121.

161. Walker, *Male Disorders of Sex,* 76; Walker later followed Freud on sexual develop-

ment, seeing homosexuality as a form of immaturity, but felt the law, which threatened the otherwise law-abiding, needed reforming. Kenneth Walker and Eric B. Strauss, *Sexual Disorders in the Male* (London: Hamish Hamilton, 1944), 156, 182.

162. Wiesner, *Sex*, 246. Wiesner ridiculed the practice of referring to the "homosexual" behavior of animals. On scientists talking about "homosexual" rats and applying their findings to humans, see also Heino Meyer-Bahlburg, "Sex Hormones and Male Homosexuality in Comparative Perspective," *Archives of Sexual Behavior* 6, no. 4 (1977): 297–326.

163. William J. Robinson, *Our Mysterious Glands* (New York: Eugenics Publishing, 1934), 146.

164. See, for example, S. J. Glass, H. J. Deuel, and C. A. Wright, "Sex Hormone Studies in Male Homosexuality," *Endocrinology* 26 (1940): 590–94; and the rejoinder of Alfred A. Kinsey, "Homosexuality: Criteria for a Hormonal Explanation of the Homosexual," *Journal of Clinical Endocrinology* 1 (May 1941): 424–28.

165. L. R. Broster, *Endocrine Man: A Study in the Surgery of Sex* (New York: Grune and Stratton, 1945), 93, 98, 104; and see also Alison Orum, *Her Husband Was a Woman! Women's Gender-Crossing in Modern British Popular Culture* (London: Routledge, 2007), 112–14.

166. L. R. Broster, "The Andrenogenital Syndrome," *Lancet* 223 (April 21, 1934): 834.

167. Theodore H. Larson, *Physicians' and Surgeons' Text-Book on Endocrinology and Ready Reference Therapy* (Los Angeles: Chicago College of Endocrinology, 1934); David Serlin, *Replaceable You: Engineering the Body in Postwar America* (Chicago: University of Chicago Press, 2004), 122.

168. Berman, *The Glands Regulating Personality*, 248–52.

169. *The New Age*, March 16, 1922, 259–61.

170. *Eugenics Review* 14 (1922–23): 194–95

171. Lipschütz, *The Internal Secretions of the Sex Glands*, 483–90.

172. See *Eugenics Review* 15 (1923–24): 327; 17 (1925–26): 110–11.

173. "Rejuvenation," *Eugenics Review* 22 (1930): 274.

174. A. F. Tredgold, "Educability and Inheritance," *Eugenics Review* 14 (1922–23): 171–72; and see also the review of J. T. Cunningham, *Hormones and Heredity* (London: Constable, 1921), in *Eugenics Review* 14 (1922–23): 126–28.

175. *Eugenics Review* 15 (1923–24): 616; 18 (1926–27): 361. On Crew's work on the Sex Hormone Committee set up by the Medical Research Council in 1930, see Alan Parkes, *Off-Beat Biologist: The Autobiography of Alan S. Parkes* (Cambridge: Galton Foundation, 1985), 195–99.

176. Arthur Walton, "'Rejuvenation' and Voronoff's Racial Claims," *Eugenics Review* 20 (1929): 253–57; and see also E. M., "Voronoff's Rejuvenation Method," *Eugenics Review* 21 (1929): 281; *Eugenics Review* 19 (1927): 142.

177. One of its last references was in a cool review of Peter Schmidt's work by William C. Miller, "Rejuvenation by Gland-Grafting," *Eugenics Review* 33 (1931): 155–57. A more positive response was made in *Eugenical News* 16 (1931): 203.

178. Edmond Székely, *Sexual Harmony and the New Eugenics*, trans. Purcell Weaver (London: C. W. Daniel, 1938); R. G. Van Yelyr, *Sex, Love, and Eugenics: A Modern Guide to Sexual Health and Happiness* (London: Gerald G. Swan, 1941).

179. On the Soviet Union, for example, see Eric Naiman, "Discourse Made Flesh: Healing and Terror in the Construction of Soviet Subjectivity," in *Language and Revolution: Making Modern Political Identities,* ed. Igal Halfin, (London: Frank Cass, 2002), 287–316; Frances L. Bernstein, "Panic, Potency and the Crisis of Nervousness in the 1920s," in *Everyday Life in Early Soviet Russia: Taking the Revolution Inside,* ed. Christina Kiaer and Eric Naiman (Bloomington: Indiana University Press, 2006), 170.

180. On the pathological rhetoric of the period, see Richard Overy, *The Morbid Age: Britain between the Wars* (London: Allen Lane, 2009).

181. Nikolas Rose, "The Politics of Life Itself," *Theory, Culture and Society* 18, no. 6 (2001): 3.

CHAPTER 5

1. Joseph Čapek and Karel Čapek, *R.U.R. and The Insect Play*, trans. P. Selver (London: Oxford University Press, 1961), 96.

2. J. B. S. Haldane, *Daedalus; or, Science and the Future* (London: Kegan Paul, Trench, Trübner, 1924). On sex determination and artificial insemination soon being possible for humans, see Enid Charles, *The Menace of Under-Population: A Biological Study of the Decline of Population Growth* (London: Watts, 1936), 192; and see also Aline Ferreira, "The Sexual Politics of Ectogenesis in the To-Day and To-Morrow Series," *Interdisciplinary Science Review* 34, no. 1 (2009): 32–55.

3. Aldous Huxley, *Brave New World* (New York: HarperPerrenial, 1998 [1932]), 5.

4. In court, Peter White and S. J. Birbaum appeared for the plaintiff, Lillian Grace Orford (née Partridge). W. R. Smyth and E. G. McMillan represented the defendant, Frederick Orford.

5. *Toronto Daily Star,* November 23, 1920, 3.

6. *Toronto Daily Star,* November 23, 1920, 3.

7. *The Globe,* December14, 1920, 8.

8. *The Globe,* December 11, 1920, 15.

9. Sir George Edmund Hodgkinson, sheriff of London (d. 1886), had two sons. In 1914, George Edmund Hodgkinson was listed as the solicitor, 124 Chancery Lane, London, of the London Film Company, Limited. *The Times,* July 30, 1914, 18. The Bloomsbury painter Mark Gertler (1891–1939) in 1930 married Marjorie Greatorex (née Hodgkinson, later Kostenz), daughter of George Edmund Hodgkinson, a London solicitor. They honeymooned with Aldous Huxley. See John Rothenstein, *Modern English Painters* (London: Taylor and Francis, 1952–74): 2, 208; John Woodson, *Mark Gertler: Biography of a Painter, 1891–1939* (London: Sidgwick and Jackson, 1972), 294.

10. Presumably W. Rayner, the author of "Cyanosis in Newly Born Children Caused by Aniline Marking Ink," *British Medical Journal* 1 (1886): 294.

11. Dr. Arthur F. Stabb, of Queen Charlotte's Hospital, was an expert in Caesarian section. See Dr. J. Barris, "The Treatment of Pregnancy Complicated by Morbus Cordis by Means of Cæsarean Section under Spinal Anæsthesia," *British Journal of Obstetrics and Gynaecology* 25, no. 4 (1914): 186–91.

12. *Orford v. Orford* (1921), *Ontario Law Reports,* 18.

13. Lillian at first denied and then later admitted going for a time under the name of Mrs. Joyce Dane (or Dehn) and was believed by some to be Captain Dane's wife. She lived for a time in Sussex in a place leased by either Dane or Hodgkinson. She also admitted to having used the name Lillian Grace Hodgkinson.

14. White, citing the *American and English Encyclopedia of Law*, in *Orford v. Orford*, 21,

15. *Orford v. Orford*, 20. Orde ignored the possibility that the use of a contraceptive would prevent one's reproductive powers from being surrendered. On adultery being committed despite a hysterectomy or old age rendering reproduction impossible, see Norman St. John-Stevas, *Life, Death and the Law: Law and Christian Morals in England and the United States* (London: Beard Books, 2002), 130–31; and see also Bernard M. Dickens, *Medico-Legal Aspects of Family Law* (Toronto: Butterworths, 1979), 3–9. Due to the Orford judgment, adultery in Canada included neither bestiality nor homosexual acts.

16. *Orford v. Orford*, 23.

17. The petition for divorce was made to the House of Commons on February 14, 1921, and assent given on June 4, 1921. *Acts of the Parliament of the Dominion of Canada* (Ottawa: Thomas Mulvey, 1921), 149. In addition to no longer having to pay alimony, Frederick Orford also freed his properties from the bar of dower, which on their marriage he had settled on Lillian. See *Re Orford and Danforth Heights Limited* (1921), *Ontario Law Reports*, 25.

18. James G. Snell, *In the Shadow of the Law: Divorce in Canada, 1900–1939* (Toronto: University of Toronto Press, 1991), 92–94; G. P. R. Tallin, "Artificial Insemination," *Canadian Bar Review* 34, no. 1 (January 1956): 166, 128.

19. Fred Rosner, *Studies in Torah Judaism: Modern Medicine and Jewish Law* (New York: Yeshiva University, 1972), 91–93; Michael J. O'Dowd and Elliot Elias Philipp, eds., *The History of Obstetrics and Gynaecology* (New York: Parthenon Publishing, 1994), 367; W. J. Finegold, *Artificial Insemination with Husband Sperm* (Springfield, IL: Charles C. Thomas, 1964), 4.

20. S. Williams, ed., *Sir Thomas Browne's Works* (London: Pickering, 1835), 3:345.

21. Irving C. Rosse, "Sexual Incapacity in Its Medico-Legal Relations," in *Medical Jurisprudence: Forensic Medicine and Toxicology*, ed. R. A. Witthaus and Tracy C. Becker (New York: William Wood, 1894), 2:387.

22. A. M. C. M. Schellen, *Artificial Insemination in the Human* (Amsterdam: Elsevier Publishing, 1957).

23. Clara Pinto-Correia, *The Ovary of Eve: Egg and Sperm and Preformation* (Chicago: University of Chicago Press, 1997), 196, 203; Cynthia R. Daniels, *Exposing Men: The Science and Politics of Male Reproduction* (New York: Oxford University Press, 2006), 21.

24. Everard Home, "An Account of the Dissection of an Hermaphrodite Dog," *Philosophical Transactions of the Royal Society* (1799): 162; and see also Angus McLaren, *Reproductive Rituals: The Perception of Fertility in England from the Sixteenth to the Nineteenth Century* (London: Methuen, 1984), 13–14.

25. "It has, hence, occurred to many philosophical minds, to overleap this barrier at once, by throwing the fructifying agent right into the cavity of the uterus." Sims succeeded in achieving only one conception, and the woman miscarried. J. Marion Sims, *Clinical Notes on Uterine Surgery* (New York: William Wood, 1866), 364.

26. In describing a range of male impediments—including inordinate size of penis, tight prepuce, hypospadias, failure of erection, and premature ejaculation—and in explaining how a doctor might instruct a husband on how to employ a syringe, Thouret appears to be following Hunter. See M. A. Thouret, *Application sur l'espèce humaine des expériences faites par Spallanzani sur quelques animaux, relativement à la fécondation artificielle des femmes ou résultats d'une expérience qui prouve que l'on peut créer des enfants avec le concours des deux sexes mais sans leur approche* (Paris: Thorens, 1803). Later texts also noted that women like Lillian Orford suffering from a retroflexed uterus or vaginismus could be assisted by artificial insemination.

27. Charles Robin, "La Fécondation," in *Dictionnaire encyclopédique des sciences médicales en plusieurs volumes,* ed. Amédée Dechambre (Paris: G. Masson et P. Asselin, 1876), 596; F. N. L. Poynter, "Hunter, Spallanzani, and the History of Artificial Insemination," in *Medicine, Science and Culture,* ed. L. G. Stevenson and R. P. Multhauf (Baltimore: Johns Hopkins University Press, 1968), 101.

28. Dr. J. Gérard, *Traité pratique des maladies de l'appareil génital de la femme, avec une notice sur la stérilité et le moyen d'y remédier par la fécondation artificielle* (Paris: V.-A. Delahaye, 1877), 382.

29. "On Artificial Fecundation," *Canada Lancet* 3, no. 2 (1870): 46–47. In this account, Girault was cited as saying that by simply employing a uterine catheter through which he blew sperm, he had successes in ten of twenty-seven cases. See also Wendy L. Mitchinson, "Historical Overview of Medical Perceptions of Infertility in Canada, 1850–1950," *The Prevalence of Infertility in Canada, Research Studies of the Royal Commission on New Reproductive Technologies* (Ottawa: Canada Communications Group, 1993), 6:1–60

30. Pierre-Fabien Gigon, *Essai sur la fécondation artificielle chez la femme dans certains cas de stérilité* (Paris: A. Parent, 1871).

31. See Oscar de Lajartre, *Des grands obstacles à la fécondation: Suppression rapide de la stérilité* (Paris: Chez l'auteur, 1888); "Artificial Fecundation," *Annals of Hygiene* (1888): 98; George P. Smith, II, "Through a Test Tube Darkly: Artificial Insemination and the Law," *Michigan Law Review* 67, no. 1 (1968): 129, citing Rice, *Notre Dame Law Review* 34 (1959): 510.

32. The case won worldwide attention. See "The Moral Bearings of Artificial Impregnation," *Boston Medical and Surgical Journal* 110 (May 1, 1884): 430. The report cited Courty, Pajot, De Sinéty, Lutaud, Eustache, Sims, and Gaillard Thomas as supporters of insemination. See also G. W. Bartholomew, "Legal Implications of Artificial Insemination," *Modern Law Review* 21 (May 1958): 238.

33. The same timing of the "rut humain" was recommended by Dr. J. Gérard, who claimed forty-one successes out of seventy-two treated. *Traité pratique des maladies de l'appareil génital de la femme,* 374, 385. The fabulous claim of 504 successes and only sixty-three failures was made in Lajartre, *Des grands obstacles à la fécondation,* 113.

34. Jules Gautier, *La fécondation artificielle et son emploi contre la stérilté chez la femme* (Paris: J. B. Baillière et fils, 1889), 57, 131–32. On overviews of artificial insemination, see also Dr. A. Coustan, "Impuissance," in *La Grande encyclopédie* (Paris: Lamirault, 1887–1902), 647–48.

35. Adele E. Clarke, *Disciplining Reproduction: Modernity, American Life Sciences, and "the Problems of Sex"* (Berkeley: University of California Press, 1998).

36. The best known was Yveling Rambaud and Dubut de Laforest, *Le Faiseur d'hommes* (Paris: C. Marpon et E. Flammarion, 1884). For the claim that a woman, who detested men but wanted a child, had herself injected by syringe in a brothel, see René Schwaeblé, *Les Détraqués de Paris, étude de mœurs contemporaines* (Paris: Darangon, 1910), 247.

37. Michael Finn, "Female Sterilization and Artificial Insemination at the French Fin-de-Siècle: Facts and Fictions," *Journal of the History of Sexuality* 18, no. 1 (2009): 39.

38. See Abel Faivre, "Alors, c'est bien vrai, madame? Vous voulez essayer de la fécondation artificielle?," in "Les Médecins," special issue, *L'Assiette au beurre,* March 22, 1902; and see the illustrations of José Roy in Dr. J. Gérard, *Nouvelles causes de stérilité dans les deux sexes: Fécondation artificielle comme moyen ultime de traitement* (Paris: C. Marpon et E. Flammarion, 1888).

39. Drs. Jaf and Saldo, *Physiologie secrète de l'homme et de la femme* (Paris: Denans, 1908), 36; Dr. Saldo, *Love without Danger, Followed with The Short Act* (Paris: Editions Modernes, 1927), 94; Dr. Caufeynon, *La Procréation à volonté des filles et des garçons, suivie de la fécondation artificielle* (Paris: P. Fort, 1903). Drs. Jaf and Caufeynon, *Sécurité des deux sexes en amour* (Paris: Georges-Anquetil, 1926), 429–30; and for the claim that the French remained the experts in "artificial impregnation," see V. S. Whitehead and C. A. Hoff, *Ethical Sex Relations; or, The New Eugenics* (Chicago: John A. Hertel, 1928), 127.

40. Bram Djkstra, *Idols of Perversity: Fantasies of Feminine Evil in Fin-de-Siècle Culture* (New York: Oxford University Press, 1986), 19.

41. Georges Vacher de Lapouge, *Sélections sociales* (Paris: A. Fontemoing, 1896), 472, 473, cited in William H. Schneider, "The Eugenics Movement in France, 1890–1940," in *The Wellborn Science: Eugenics in Germany, France, Brazil, and Russia,* ed. Mark B. Adams (New York: Oxford University Press, 1990), 71; and see also Martin Richards, "Artificial Insemination and Eugenics: Celibate Motherhood, Eutelegenesis and Germinal Choice," *Studies in History and Philosophy of Biological and Biomedical Sciences* 39, no. 2 (June 2008): 211. Lapouge also claimed that he performed the world's first case of "telegenesis" by sending a vial of human sperm from Béziers to Montpellier, where it was successfully employed. Jennifer Michael Hecht, "The Solvency of Metaphysics: The Debate over Racial Science and Moral Philosophy in France, 1890–1919," *Isis* 90, no. 1 (1999): 6n9.

42. Gerald Kelley, S.J., "The Morality of Artificial Insemination," *Ecclesiastical Review* 101 (1939): 109–18. Note that Louis Girault, *Étude sur la génération artificielle dans l'espèce humaine* (Paris: Aux bureaux de l'Abeille médicale, 1869), was placed on the Vatican's list of suspect books; see Hubert Wolf, *Index* (Munich: C. H. Beck, 2006), 275.

43. Theodore van de Velde, *Fertility and Sterility in Marriage: Their Voluntary Promotion and Limitation,* trans. F. W. Stella Browne (London: W. Heinemann, 1931), 272–73; G. W. Bartholomew, "The Development and Use of Artificial Insemination," *Eugenics Review* 49 (1958): 188.

44. See, for example, Jean De Bonnefon, *Les Cas de conscience modernes* (Paris: Ambert, 1904), 106–7; William Kevin Glover, *Artificial Insemination among Human Beings: Medical, Legal and Moral Aspects* (Washington, DC: Catholic University of America Press, 1948).

45. Paolo Mantegazza, *Sullo sperma umano* (Rendiconti: Reale Istituto lombardo di scienze e lettere, 1866), cited in Gary N. Clarke, "A.R.T. and History, 1678–1978," *Human Reproduction* 21, no. 7 (2006): 1645–50; Poynter, "Hunter, Spallanzani, and the History of Artificial Insemination," 104.

46. As a medical student, Dr. Hard viewed the first artificial insemination by donor in America, carried out in 1884 by Dr. William Pancoast at Jefferson Medical College. A. D. Hard, "Artificial Impregnation," *Medical World* 27 (1909): 163–64, 306, cited in David Horn, "Unnatural Acts: Procreation and the Genealogy of Artifice," in *Processed Lives: Gender and Technology in Everyday Life,* ed. Jennifer Terry and Melodie Calvert (London: Routledge, 1997), 106–7. And see also Elaine Tyler May, *Barren in the Promised Land: Childless Americans and the Pursuit of Happiness* (New York: HarperCollins, 1995), 65–67.

47. R. L. Dickinson, "Artificial Impregnation; Essays in Tubal Insemination," *American Journal of Obstetrics and Gynecology* 1 (1920): 255–26, 306–9; and see also G. Frank Lydston, "Artificial Fertilization," *Journal of the American Medical Association* 75 (1920): 193; Jennifer Terry, *An American Obsession: Science, Medicine, and Homosexuality in Modern Society* (Chicago: University of Chicago Press, 1999), 436nn72–75; Ian R. Dowbiggin, *The Sterilization Movement and Global Fertility in the Twentieth Century* (New York: Oxford University Press, 2008), 55–56.

48. See, for example, Brooke M. Anspach, *Gynecology* (Philadelphia: J. B. Lippincott, 1921), 606–7.

49. *New York Times*, June 13, 1912, 4.

50. Christina Beninghaus, "Great Expectations—German Debates about Artificial Insemination around 1912," *Studies in History and Philosophy of Biological and Biomedical Sciences* 38 (2007): 374–92.

51. Glenn Ramsey, "The Rites of Artgenossen: Contesting Homosexual Political Culture in Weimar Germany," *Journal of the History of Sexuality* 17 (2008): 93; and on sperm collection, see Magnus Hirschfeld, *Sexual Pathology: A Study of Derangements of the Sexual Instinct,* trans. Jerome Gibbs (New York: Emerson Books, 1947), 3:351.

52. Walter Heape, "The Artificial Insemination of Mammals and Subsequent Possible Fertilisation or Impregnation of their Ova," *Proceedings of the Royal Society of London* 61B (1897): 56.

53. "Artificial Fecundation before the Inquisition," *British Medical Journal* 1 (March 5, 1898): 644. Gynecology textbooks made brief references to the subject. Eden and Lockyer, describing how a syringe could be used to take semen immediately from the vagina and insert it into uterus, asserted that few women would choose such an indelicate and chancy procedure. Thomas Watts Eden and Cuthbert Lockyer, *Gynaecology for Students and Practitioners* (London: Churchill, 1916), 150–51. The same point that only as a last resort would desperate women submit to such an undertaking was made in Ten Teachers, under the direction of Comyns Berkeley, *Diseases of Women* (London: Arnold, 1919), 109.

54. George Ernest Herman, *Diseases of Women* (London: Cassell, 1907), 647.

55. Schellen, *Artificial Insemination in the Human*, 22.

56. Havelock Ellis, *Studies in the Psychology of Sex* (Philadelphia: F. A. Davis, 1901–28), 6:632.

57. Paul Peppis, "Rewriting Sex: Mina Loy, Marie Stopes, and Sexology," *Modernism/modernity* 9, no. 4 (2002): 561–79. On the public discussion of female sexuality being legitimated in the 1920s, see Billie Melman, *Women and the Popular Imagination in the Twenties: Flappers and Nymphs* (Basingstoke: Macmillan, 1988).

58. Katherine Holden, "'Nature Takes No Notice of Morality': Singleness and Married Love in Interwar Britain," *Women's History Review* 11, no. 3 (2002): 495.

59. Marie Carmichael Stopes, *Married Love: A New Contribution to the Solution of Sex Difficulties* (London: G. P. Putnam's Sons, 1919), 151–52; and for a eugenic defense of such a policy, see Seth K. Humphrey, *The Racial Prospect* (New York: Charles Scribner's Sons, 1920), 251; Charles Wickstead Armstrong, *The Survival of the Unfittest* (London: C. W. Daniel, 1927), 100.

60. Diana H. Wyndham, "Striving for National Fitness: Eugenics in Australia, 1910s to 1930s" (PhD diss., History Department, University of Sydney, 1996), 72–73, 72–90, 151–52; and see Richards, "Artificial Insemination and Eugenics," 212; Ann Curthoys, "Piddington, Marion Louisa (1869–1950)," *Australian Dictionary of Biography* (Melbourne: Melbourne University Press, 1988), 11:226–27; Ann Curthoys, "Eugenics, Feminism and Birth Control: The Case of Marion Piddington," *Hecate* 15, no. 1 (1989): 73–89.

61. Marie Carmichael Stopes, *Married Love; or, Love in Marriage,* with preface and notes by William J. Robinson, MD (New York: The Critic and Guide, 1918), 136–37.

62. *Journal of Social Hygiene* 6, no. 2 (1920): 292–93.

63. Halliday G. Sutherland, *Birth Control: A Statement of Christian Doctrine against the Neo-Malthusians* (London: Hardy and More, 1922).

64. *The Times,* February 24, 1923, 5.

65. Martin Richards, "Genes, Genealogies and Paternity: Making Babies in the Twenty-First Century," in *Freedom and Responsibility in Reproductive Choice,* ed. J. R. Spencer and Antje du-Bois Pedain (Oxford: Hart Publishing, 2006), 56.

66. On the Margaret Sanger Research Bureau (under Abraham Stone), see Sophia Smith Collection, Margaret Sanger Papers, 1761–1995, Artificial insemination, 1941–56, Box 73: folder 7.

67. Frank M. C. Forster, "Haire, Norman" (1892–1952)," *Australian Dictionary of Biography* (Melbourne: Melbourne University Press, 1996), 14:353–54.

68. Norman Haire, *Hymen; or, The Future of Marriage* (London: Kegan Paul, Trench, Trübner, 1927), 76.

69. Haire, *Hymen; or, The Future of Marriage*, 87.

70. Norman Haire, "Artificial Insemination," *Journal of Sex Education* 1 (1948): 104; and see also Norman Haire, *Everyday Sex Problems* (London: Frederick Muller, 1949); Alison Bashford and Carolyn Strange, "Public Pedagogy: Sex Education and Mass Communication in the Mid-Twentieth Century," *Journal of the History of Sexuality* 13 (2004): 56.

71. That Aarons was otherwise morally conservative is suggested by his bleak portrayal of youthful masturbation. See S. Jervois Aarons, *Gynaecological Therapeutics* (London: Baillière, Tindall and Cox, 1910), 172–74.

72. Hugo Vickers, *Elizabeth the Queen Mother* (London: Hutchison, 2006), 97. Cooper claimed that Stephen Tennant (b. 1906), the epitome of the 1920s gay Bright Young Thing,

was so conceived. His parents' marriage was fragile, but his biographer asserts there is no evidence to support Cooper's assertions. Philip Hoare, *Serious Pleasures: The Life of Stephen Tennant* (London: Penguin, 1991).

73. "A Society" first appeared in *Monday or Tuesday* (1921). See *The Complete Shorter Fiction of Virginia Woolf* (London: Hogarth Press, 1985), 124–25; and see also Jane Garrity, *Step-Daughters of England: British Women Modernists and the National Imaginary* (Manchester: Manchester University Press, 2003), 55.

74. D. H. Lawrence, *Lady Chatterley's Lover* (New York: Grove Press, 1982 [1928]), 115, 116, 199, 223.

75. Anthony Ludovici, *Lysistrata; or, Woman's Future and Future Woman* (London: Kegan Paul, Trench, Trübner, 1924), 90.

76. See Georges Duhamel, *Scènes de la vie future* (Paris: Mercvre de France, 1956 [ca. 1930]), 38, which appeared in England as *America the Menace: Scenes from the Life of the Future*, trans. Charles Miner Thompson (London: George Allen and Unwin, 1930).

77. *The Times,* December 27, 1930, 8.

78. Lesley A. Hall, "Desperately Seeking Stella: The Pleasures and Problems of Doing Women's Biography," http://homepages.primex.co.uk/~lesleyah/lwhntlk.htm, accessed December 19, 2010; and see also Lesley A. Hall, *The Life and Times of Stella Browne: Feminist and Free Spirit* (London: I. B. Tauris, 2011).

79. Michael Tippett, *The Twentieth-Century Blues: An Autobiography* (London: Hutchinson, 1991), 56.

80. Van de Velde, *Fertility and Sterility in Marriage*, 259.

81. Van de Velde, *Fertility and Sterility in Marriage,* 262–67. The same scenario was followed in H. Rohleder, *Test Tube Babies: A History of Artificial Impregnation of Human Beings* (New York: Panurge Press, 1934), a translation of *Die Künstliche Zeugung (Befruchtung) im Tierreich* (Leipzig: Georg Thieme, 1921).

82. Magnus Hirschfeld, *Geschlechtskunde auf Grund dreißigjähriger Forschung und Erfahrung bearbeitet* (Stuttgart: Julius Püttmann, Verlagsbuchhandlung, 1926–1930), 2:407, cited in van de Velde, *Fertility and Sterility in Marriage*, 270.

83. *Daily Mirror*, November 9, 1936, 2. And see Samuel Meaker, *Human Sterility* (Baltimore: Williams and Wilkins, 1934); Margaret Marsh and Wanda Ronner, *The Fertility Doctor: John Rock and the Reproductive Revolution* (Baltimore: Johns Hopkins University Press, 2008), 108–22.

84. *Daily Mirror*, July 17, 1939, 1; see also July 29, 1943, 8.

85. Edward F. Griffith, *The Childless Family: Its Causes and Cure* (London: Kegan Paul, Trench, Trübner, 1939), 119.

86. George Ryley Scott, *Three Hundred Sex, Marriage and Birth Control Questions Answered* (London: T. Werner Laurie, 1941), 138.

87. See Arthur Giles, "The Causes, Prognosis and Treatment of Steriiity," *British Journal of Obstetrics and Gynaecology* 28, no. 2 (1921): 250; "Review of Current Literature," *British Journal of Obstetrics and Gynaecology* 30, no. 2 (1923): 300; "Review of Current Literature," *British Journal of Obstetrics and Gynaecology* 31, no. 3 (1924): 503; "Review of Current Literature," *British Journal of Obstetrics and Gynaecology* 41, no. 3 (1934): 462; "Artificial Insemi-

nation," *Lancet* 2 (1936): 1357; "Review of Current Literature," *British Journal of Obstetrics and Gynaecology* 45, no. 2 (1938): 329–69; "Review of Current Literature," *British Journal of Obstetrics and Gynaecology* 46, no. 1 (1939): 167; "Review of Current Literature," *British Journal of Obstetrics and Gynaecology* 46, no. 2 (1939): 334; "Artificial Insemination," *Lancet* 1 (1940): 203–4.

88. Naomi Pfeffer, "Artificial Insemination, In-Vitro Fertilization and the Stigma of Infertility," in *Reproductive Technologies: Gender, Motherhood, and Medicine,* ed. Michelle Stanworth (Cambridge: Polity Press, 1987), 92–95.

89. A Tory MP, raised to the Lords in 1942, John Moore-Brabazon, First Baron Brabazon of Tara, in 1910 had become the first person in the United Kingdom to qualify as an airplane pilot.

90. "Artificial Insemination," *Medico-Legal Journal* 11 (July–September 1943): 166–67; *The Times*, July 29, 1943, 8. And see also *The Times*, July 31, 1943, 5; "Artificial Insemination," *Lancet* 242 (August 7, 1943): 168.

91. According to the first British report on artificial insemination coauthored by Barton, thirty attempts of artificial insemination by husband resulted in twenty-one clear failures, nine conceptions, and four births. Fifteen attempts at artificial insemination by donor resulted in ten conceptions and eight births. Mary Barton, Kenneth Walker, and B. P. Wiesner, "Artificial Insemination," *British Medical Journal* 1 (1945): 40–43.

92. Mary Barton's husband and colleague Bertold Paul Wiesner (1901–1972), an Austrian-born sex physiologist, was one of Britain's pioneering sex researchers. In the previous chapter, his work on pregnancy diagnosis was noted. Wiesner was involved in the birth control movement, supporting the British Social Hygiene Council rather than the Family Planning Association. In the mid-1930s, his sex hormone grants ended, and Wiesner came to London where he was attached to the Royal Northern Hospital as a biologist while also running the capital's only seminology lab and assisting at Mary Barton's infertility clinic. See Cedric Lane-Roberts, Albert Sharman, Kenneth Walker, and B. P. Wiesner, *Sterility and Impaired Fertility* (London: Hamish Hamilton, 1939); Merriley Borell, "Biologists and the Promotion of Birth Control Research, 1918–1938," *Journal of the History of Biology*, 20, no. 1 (1987): 51–87.

93. Mary Barton, "Artificial Insemination," *British Medical Journal* 2 (1943): 312–13; and see also Cedric Lane Roberts, Albert Sharman, Kenneth Walker, B. P. Wiesner, and Mary Barton, *Sterility and Impaired Fertility, Pathogenesis, Diagnosis and Treatment* (London: Hamish Hamilton, 1949).

94. Cheryl Law, *Suffrage and Power: The Women's Movement, 1918–1928* (London: I. B. Tauris, 1997), 170; Pauline Mazumdar, *Eugenics, Human Genetics, and Human Failings: The Eugenics Society, Its Sources and Its Critics in Britain* (London: Routledge, 1992) 201, 211, 310n43.

95. For her hostility to contraception, see Letitia Fairfield, "The State and Birth Control," in *Medical Views on Birth Control,* ed. Sir James Marchant (London: Martin Hopkinson, 1926), 69–88; and Letitia Fairfield, "Birth Control: Arguments Against," in *The Family Book: A Comprehensive Guide to Family Life from Before Marriage to the Adolescence of Children. Primarily for Parents,* ed. Gwen St. Aubyn (London: Arthur Barker,

1935). Her position was lauded by the eugenicist and anti-Semite Anthony Ludovici in *The Night-Hoers; or, The Case against Birth Control and an Alternative* (London: Herbert Jenkins, 1928), 140.

96. Letitia Fairfield, *The Case against Sterilization* (London: Catholic Truth Society, 1935); Letitia Fairfield, *Catholics and the German Law of Sterilization* (London: Burns, Oaks and Washbourne, 1938).

97. "Pitfalls in the Practice of Obstetrics and Gynaecology," *Medico-Legal Journal* 11 (April–June 1943): 80. Just such a case occurred in L. v. L. [1949] 1 All E.R. 141. See "Nullity—Incapacity—Artificial Insemination," *Medico-Legal Journal* 17 (1949): 33–34.

98. "Pitfalls in the Practice of Obstetrics and Gynaecology," 81; and see also "Artificial Insemination," *British Medical Journal* 1 (1945): 96; "Artificial Insemination of Women," *Medico-Legal Journal* 13 (April–June 1945): 114.

99. Robert Forbes, "The Medico-Legal Aspects of Artificial Insemination," *Medico-Legal Journal* 12 (July–September 1944): 147–48. Orford was also cited in Haire, "Artificial Insemination," 106; William Kevin Glover, *Artificial Insemination among Human Beings: Medical, Legal and Moral Aspects* (Washington, DC: Catholic University of America Press, 1948), 45–48. See also Cecil Binney, "Legal and Social Implications of Artificial Insemination," *Eugenics Review* 40 (1948–49): 199–204.

100. In the famous Russell vs. Russell divorce case, Lord Dunedin declared that a woman who had not been penetrated but had sex ab extra and conceived a child by a man other than her husband was guilty of adultery. Law Reports, Appeal Cases, 1924; House of Lords, Russell v. Russell [1924] AC 687, 721, 726; and see also Eileen Hunter, *Christabel: The Russell Case and After* (London: Deutsch, 1973).

101. "Pitfalls in the Practice of Obstetrics and Gynaecology," 81.

102. "Artificial Insemination," *British Medical Journal* 1 (1945): 236.

103. Forbes, "The Medico-Legal Aspects of Artificial Insemination," 148; "Artificial Insemination," *British Medical Journal* 2 (1944): 672. Forbes was secretary of the Medical Defence Union.

104. "Artificial Insemination," *British Medical Journal* 1 (1947): 605–6.

105. Binney, "Legal and Social Implications of Artificial Insemination," 199; "The technique is surprisingly simple," asserted Wilfred Shaw, *Textbook of Gynaecology* (London: J. and A. Churchill, 1948), 277.

106. Emil Novak, *The Woman Asks the Doctor* (London: Christophers, 1935), 157.

107. Angus McLaren, *Impotence: A Cultural History* (Chicago: University of Chicago Press, 2007), 183.

108. American women learned of a range of therapies including artificial insemination, thanks to radical practitioners like Edward Bliss Foote who sold syringes by mail. Margaret Marsh and Wanda Ronner, *The Empty Cradle: Infertility in America from Colonial Times to the Present* (Baltimore: Johns Hopkins University Press, 1996), 69–70.

109. Pfeffer, "Artificial Insemination, In-Vitro Fertilization and the Stigma of Infertility," 86.

110. F. H. Stevenson, "Artificial Insemination," *British Medical Journal* 2 (1944): 642.

111. *New York Times*, April 9, 1945, 3.

112. One inventive physician suggested that if a husband were impotent a Coitus Training Apparatus (a contraption that supported the penis) might restore his potency. "Mechanotherapy" could thus replace the need of both masturbation and artificial insemination. Joseph Loewenstein, "An Alternative to Artificial Insemination," *Journal of Sex Education* 1 (1948): 92–95.

113. H. P. Newsholme, "Artificial Insemination," *British Medical Journal* 2 (1944): 641–42; and see H. P. Newsholme, *Christian Ethics and Social Health* (London: John Heritage, Unicorn Press, 1937).

114. "Artificial Insemination," *British Medical Journal* 1 (1944): 95.

115. "Artificial Insemination," *British Medical Journal* 1 (1945): 165.

116. "Artificial Insemination," *British Medical Journal* 1 (1945): 165.

117. "Artificial Insemination," *British Medical Journal* 1 (1945): 165.

118. "Artificial Insemination," *British Medical Journal* 1 (1945): 236.

119. "Artificial Insemination," *British Medical Journal* 1 (1944): 25.

120. "Artificial Insemination," *British Medical Journal* 1 (1945): 347.

121. R. M. Ladell, "Artificial Insemination," *British Medical Journal* 2 (1944): 707. R. Macdonald Ladell was author of *The Sex Education of Children* (Birmingham: Cornish Bros., 1934).

122. "Artificial Insemination," *British Medical Journal* 1 (1945): 236

123. Kenneth Walker, "Artificial Insemination," in *Problems of Fertility in General Practice*, ed. Margaret Hadley Jackson, Joan Malleson, John Stallworthy, and Kenneth Walker (London: H. Hamilton Medical Books, 1948), 119; and see also Haire, "Artificial Insemination," 58–62.

124. "Artificial Insemination," *British Medical Journal* 1 (1947): 738–39.

125. "Artificial Insemination," *British Medical Journal* 1 (1945): 236.

126. Margaret H. Jackson, "Artificial Insemination," *British Medical Journal* 2 (1944) 577–78. The Exeter Women's Welfare Association clinic was established in 1930. Around 1937, some couples began asking for advice on infertility. By 1943, one-third of patients were seeking such information. Margaret Hadley Jackson, "A Medical Service for the Treatment of Involuntary Sterility," *Eugenics Review* 36 (1944–45): 117–25.

127. "Artificial Insemination," *British Medical Journal* 1 (1947): 945; and see also Margaret Hadley Jackson, "Artificial Insemination (Donor)," *Eugenics Review* 48 (1957): 203–7.

128. "Artificial Insemination," *British Medical Journal* 2 (1943): 434. The sexual misery of unmarried women was a common trope in the interwar works of progressive writers like Bertrand Russell, Havelock Ellis, and George Orwell. See Katherine Holden, *The Shadow of Marriage: Singleness in England, 1914–1960* (Manchester: Manchester University Press, 2007).

129. Judge William George Earengey, husband of the suffragist Florence Earengey (née How and sister of birth controller Edith How-Martyn), interested himself in a range of progressive issues. See W. G. Earengey, *Woman under the Law* (London: Women's Freedom League, 1908); W. G. Earengey, "Voluntary Euthanasia," *Medico-Legal Review,* 8 (1940): 93.

130. Forbes, "The Medico-Legal Aspects of Artificial Insemination," 151.

131. "Artificial Insemination," *British Medical Journal* 1 (1945): 130.

132. Forbes, "The Medico-Legal Aspects of Artificial Insemination," 151; and for similar views, see Audrey Roberts, "Artificial Insemination," *British Medical Journal* 1 (1945): 199.

133. In the United States, Frances Seymour provided unmarried middle-class women with donor insemination, in her view "a decent and moral method of acquiring the children nature intended them to bear." Marsh and Ronner, *The Empty Cradle*, 165.

134. "Artificial Insemination (Donor)," *Eugenics Review* 48 (1957): 209–11.

135. Joan Malleson, "Artificial Insemination," *British Medical Journal* 2 (1944): 706–7.

136. "Artificial Insemination," *British Medical Journal* 1 (1945): 277. On infertility as still being regarded in the 1990s as "a medically liminal category" and the infertile as "driven" or obsessed, see Margarete Sandelowski, *With Child in Mind: Studies of the Personal Encounter with Infertility* (Philadelphia: University of Pennsylvania Press, 1993), 45.

137. In 1938, the *American Mercury* foresaw artificial insemination being used out of spite by feminists. Cynthia R. Daniels and Janet Golden, "Procreative Compounds: Popular Eugenics, Artificial Insemination and the Rise of the American Sperm Banking Industry," *Journal of Social History,* 38, no. 1 (2004): 5–6. In France, Simone de Beauvoir was to herald artificial insemination as promising to complete the control of reproduction began by contraception and abortion. Simone de Beauvoir, *The Second Sex,* trans. H. M. Parshley (New York: Knopf, 1957 [1949]), 492.

138. *The Times,* June 20, 1945, 2; June 28, 1945, 2; July 4, 1945, 2. Given the restrictions on donors, the defenders of artificial insemination asserted that it posed no greater danger of resulting in incest than did adoption. Mary Barton, Kenneth Walker, and B. P. Wiesner, "Artificial Insemination," *British Medical Journal* 1 (1945): 40–43.

139. Malleson gave birth control instructions at University College Hospital and was central in the Bourne case of 1938 that strengthened doctors' right to provide abortions on mental health grounds. See Joan Malleson, *Any Wife and Any Husband* (New York: Random House, 1952.)

140. Beatrice M. Willmott Dobbie, *Obstetrics and Gynaecology* (London: H. K. Lewis, 1948), 258.

141. *The Times*, April 20, 1945, 8; *New York Times,* April 20, 1945, 13. On Baroness Summerskill—a doctor, a supporter of Stopes's birth control campaign, and in 1938 the first woman MP not to use her married name—see Edith Summerskill, *A Woman's World* (London: Heinemann, 1967); and Cheryl Law, ed., *Women: A Modern Political Dictionary* (London: I. B. Tauris, 2000), 141–42. Her publications included *Babies without Tears* (London: Hutchinson, 1941) and *Wanted—Babies: A Trenchant Examination of a Grave National Problem* (London: Gainsborough Press, 1943). In 2003, her grandson Ben Summerskill became head of Stonewall, a British gay lobbying group.

142. Such opponents would have been further infuriated if they had known that Mary Barton's method of selecting donors was ethically questionable. Pfeffer notes that Barton's husband, Bertold Wiesner—once highly respected by the Medical Research Council and the Royal College of Obstetrics and Gynecology—was for some unknown reason by 1946 regarded as persona non grata. "I have not been able to find out what he did to be considered disreputable." Naomi Pfeffer, *The Stork and the Syringe: A Political History of*

Reproductive Medicine (Cambridge: Polity Press, 1993), 195n66. The probable reason was finally brought to light in Barry Stevens's 2009 documentary "Bio-Dad" that revealed that Wiesner secretly used his own sperm in the infertility clinic that he ran with Barton. His many offspring included Barry Stevens and David Gollancz. "Bio-Dad," *CBC Doc Zone*, February 26, 2009, on CBC-TV.

143. "Artificial Insemination," *British Medical Journal* 1 (1945): 277

144. "Artificial Insemination," *British Medical Journal* 2 (1944): 578.

145. "Artificial Insemination," *British Medical Journal* 1 (1945): 165.

146. "Artificial Insemination," *British Medical Journal* 2 (1944): 512–13.

147. "Artificial Insemination," *British Medical Journal* 1 (1947): 236.

148. "Artificial Insemination," *British Medical Journal* 1 (1947): 945.

149. Joan Malleson, "Artificial Insemination," *British Medical Journal* 2 (1944), 706–7.

150. *The Globe*, November 23, 1920, 8. See also *Toronto World*, November 23, 1920, 2.

151. *Toronto Daily Star*, November 29, 1920, 3.

152. Forbes, "The Medico-Legal Aspects of Artificial Insemination," 150.

153. See, for example, Max Hühner, *Sterility in the Male and Female and Its Treatment* (New York: Rebman, 1913), 78.

154. Mary Barton, "Artificial Insemination," *British Medical Journal* 2 (1943): 312–13.

155. In the United States, G. Frank Lydston went so far as to suggest injecting donor sperm into the vas of the husband who could decide himself whether or not to inform his wife of the medical assistance he had received. *Journal of the American Medical Association*, 75 (1920): 193.

156. On the mixing of the sperm of the donor with that of the husband so the latter might believe the conception was his (just as members of firing squads did not know if they were firing blanks), see Pfeffer, *The Stork and the Syringe*, 114.

157. Margaret H. Jackson, "Artificial Insemination," *British Medical Journal* 2 (1944) 577–78.

158. Joan Malleson, "Artificial Insemination," *British Medical Journal* 2 (1944): 706–7.

159. A. J. Brock, "Artificial Insemination, *British Medical Journal* 1 (1945): 96–97.

160. "Ersatz Fathers for Baby Nazis," *Daily Mirror*, February 1, 1940, 16; "U.S. Board Bares Atrocity Details," *New York Times*, November 26, 1944, 1, 24; and see also *New York Times*, February 1, 1940, 6; September 22, 1945, 5; September 26, 1945, 8.

161. George Orwell, *Nineteen Eighty-Four* (New York: Harcourt, Brace and World, 1949), 66.

162. "Artificial Insemination," *British Medical Journal* 2 (1943): 404

163. Audrey Roberts, "Artificial Insemination," *British Medical Journal* 1 (1945): 199.

164. "Artificial Insemination," *British Medical Journal* 1 (1945): 165.

165. "Artificial Insemination," *British Medical Journal* 1 (1945): 236.

166. "Artificial Insemination," *British Medical Journal* 2 (1943): 404.

167. "Artificial Insemination," *British Medical Journal* 2 (1943): 404.

168. "Artificial Insemination," *Lancet* 242 (August 7, 1943): 168; and see also "Case for Eugenics," *New York Times*, May 14, 1944, E9; Alfred Tyrer, *Sex, Marriage and Birth Control* (Toronto: Marriage Welfare Bureau, 1943 [1936]), 238–39; *Scientific and Technical Societies*

of the United States and Canada (Washington, DC: National Academy of Sciences, National Research Council, 1955–68), 8:217.

169. Seymour's activities inspired George Weiss to produce the American exploitation film *Test Tube Babies* (1948). See Eric Schaefer, *"Bold! Daring! Shocking! True!": A History of Exploitation Films, 1919–1959* (Durham, NC: Duke University Press, 1999), 206–8.

170. "Substitute Fathers," *Newsweek*, September 3, 1943, 87–88; Kathleen M. Woodward, *Figuring Age: Women, Bodies, Generations* (Bloomington: Indiana University Press, 1998), 105n4; Daniels and Golden, "Procreative Compounds," 5–27; Eugenics Society Collection, SA/EUG/D6 in the Wellcome Contemporary Medical Archives Centre.

171. Mary Barton, "Artificial Insemination," *British Medical Journal* 2 (1943): 313.

172. Forbes, "The Medico-Legal Aspects of Artificial Insemination," 141–42, 149.

173. "Artificial Insemination," *British Medical Journal* 1 (1945): 130.

174. A paper delivered to the 1921 International Eugenics Congress noted how in the United States artificial insemination allowed for "preferential mate selection." "The author is reliably informed that in one of the largest American cities many women whose husbands are sterile now resort to a certain physician who inseminates them artificially. As is natural, these women stipulate the race of the father whose identity they rarely know. In some cases, they are further concerned with his quality and attributes. It is obvious that this physician has a very heavy responsibility in thus determining the quality of many of these offspring. If he will, he can render a very valuable eugenic service by declining to inseminate inferior women and by obtaining the best fathers possible." Roswell H. Johnson, "Mate Selection," *Eugenics Review* 14–15 (1922–24): 262.

175. Herbert Brewer, "Eutelegenesis," *Eugenics Review* 27 (1935–36): 121–26; Daniel Kevles, *In the Name of Eugenics: Genetics and the Uses of Human Heredity* (Berkeley: University of California Press, 1985), 188–91.

176. Cyril Bibby, *Heredity, Eugenics and Social Progress* (London: Victor Gollancz, 1939), 78. On the suggestion that impregnations be carried out in doctors' offices or at matrimonial bureaux, see F. William Inman, *Biological Politics: An Aid to Clear Thinking* (London: Simpkin Marshall, 1935), 153–54.

177. H. J. Muller, *Out of the Night: A Biologist's View of the Future* (New York: Vanguard Press, 1935), 111; and see also Richards, "Artificial Insemination and Eugenics," 211–21; David Plotz, *The Genius Factory: Unravelling the Mystery of the Nobel Prize Sperm Bank* (London: Simon and Schuster, 2005).

178. Julian Huxley, *Man Stands Alone* (New York: Harper and Brothers, 1941), 77. And for the portrayal of a utopia in which couples choose frozen sperm from a central bank of superior stock, see Aldous Huxley, *Island* (London: Granada, 1976 [1962]), 219–20.

179. Walker stated that artificial insemination by donor was advocated for sterility cases, where the wife was Rh-negative, where the husband had a hereditary disability, and "as a long-term eugenic means of improving stock by inseminating with sperms from sires of outstanding mental and physical attributes." Only the first three criteria were followed in Britain. Kenneth Walker, "Artificial Insemination," in *Problems of Fertility in General Practice*, ed. Margaret Hadley Jackson, Joan Malleson, John Stallworthy, and Kenneth Walker (London: H. Hamilton Medical Books, 1948), 110.

180. The authors also asserted that some Jewish patients sought non-Jewish donors to spare the subsequent child anti-Semitic slurs. Mary Barton, Kenneth Walker, and B. P. Wiesner, "Artificial Insemination," *British Medical Journal* 1 (1945): 42. Anti-Semitism was rife in postwar Britain. In discussing donor insemination, one commentator raised the specter of "the blonde Nordic lady who feels miserable and humiliated at having a child of unmistakable Jewish paternity." Binney, "Legal and Social Implications of Artificial Insemination," 204.

181. Margaret H. Jackson, "Artificial Insemination," *British Medical Journal* 2 (1944): 577–78.

182. "Artificial Insemination," *British Medical Journal* 1 (1945): 346–47.

183. In 1952, the Americans Hannah M. Stone and Abraham Stone noted that a sterile woman sought their aid in finding a surrogate whom her husband would artificially inseminate. Hannah M. Stone and Abraham Stone, *A Marriage Manual: A Practical Guide-Book to Sex and Marriage* (London: Victor Gollancz, 1952), 174–75; and see also Laura Mamo, *Queering Reproduction: Achieving Pregnancy in the Age of Technoscience* (Durham, NC: Duke University Press, 2007).

CHAPTER 6

1. Aldous Huxley, *Brave New World* (New York: HarperPerrenial, 1998 [1932]), 23.

2. *New Statesman*, February 6, 1932, 174.

3. Raymond Williams, *The Country and the City* (London: Chatto and Windus, 1973).

4. Charles E. Rosenberg, "Pathologies of Progress: The Idea of Civilization as Risk," *Bulletin of the History of Medicine* 72, no. 4 (1998): 714–30.

5. See Ruggles Gates Papers, King College, London; *Dictionary of National Biography, 1961–1970* (Oxford: Oxford University Press, 1981), 424–25; Elazar Barkan, *The Retreat of Scientific Racism* (Cambridge: Cambridge University Press, 1992), 168–76.

6. October 16, 1931, Gates Papers, Correspondence, 2/1/5; see also Gates Papers 2/1/4. Men of the Trees circular, March 25, 1935.

7. On Ursula Grant Duff (née Lubbock), the daughter of Lord Avebury, see Arthur Keith, *Man* 34 (April 1934): 49–51; Stephen Trombley, *The Right to Reproduce: A History of Coercive Sterilization* (London: Weidenfeld and Nicolson, 1988), 117–18, 121, 140; Richard A. Soloway, *Demography and Degeneration: Eugenics and the Declining Birthrate in Twentieth-Century Britain* (Chapel Hill: University of North Carolina Press, 1990), 304, 308.

8. His papers are held at the University of Saskatchewan, Richard St. Barbe Baker fonds (hereafter Baker Papers), MG 71.

9. Richard St. Barbe Baker, *I Planted Trees* (London: Lutterworth Press, 1944), 10.

10. Richard St. Barbe Baker, *Green Glory: The Story of the Forests of the World* (London: Lutterworth Press, 1948), 180.

11. See W. Dawson, "Science in Forestry," in *Science and the Nation*, ed. A. C. Seward (Cambridge: Cambridge University Press, 1917), 129–45.

12. Caroline Ford, "Reforestation, Landscape Conservation, and the Anxieties of Empire in French Colonial Algeria," *American Historical Review* 113, no. 2 (April 2008): 341–62; Car-

oline Ford, "Nature, Culture, and Conservation in France and Her Colonies, 1840–1940," *Past and Present* 183 (May 2004): 173–98; David Blackbourn, *The Conquest of Nature: Water, Landscape, and the Making of Modern Germany* (London: Jonathan Cape, 2006).

13. "The imperial agricultural and forestry services provided an appropriate space in which to nurture dissident organic science, distant from central orthodoxy." David Matless, *Landscape and Englishness* (London: Reaktion Books, 1998), 111; and see also Gregory Allen Barton, *Empire Forestry and the Origins of Environmentalism* (Cambridge: Cambridge University Press, 2002); Peder Anker, *Imperial Ecology: Environmental Order in the British Empire* (Cambridge, MA: Harvard University Press, 2001), 7–40.

14. Richard St. Barbe Baker, *Men of the Trees: In the Mahogany Forests of Kenya and Nigeria* (New York: Dial Press, 1931); and on memories of his work in Kenya, see Wangari Maathai, *Unbowed: A Memoir* (New York; Random House, 2007), 131–32.

15. Baker, *Green Glory,* 51; and see also David M. Anderson, *Eroding the Commons: The Politics of Ecology in Baringo, Kenya, 1890s–1963* (Athens: Ohio University Press, 2002).

16. Baker, *Green Glory*, 57.

17. Philip Gibbs, *People of Destiny: Americans as I Saw Them at Home and Abroad* (New York, London: Harper and Brothers, 1920). One British science-fiction writer imagined a future Americanized world state in which sacred rapes and lynchings were celebrated in memory of the United States' past. Olaf Stapledon, *Last and First Men: A Story of the Near and Far Future* (New York: Dover, 1931).

18. C. E. M. Joad, *Thrasymachus; or, The Future of Morals* (New York: E. P. Dutton, 1926), 29.

19. Aldous Huxley, "America," *Jesting Pilate* (1926), in *Complete Essays,* ed. Robert S. Baker and James Sexton (Chicago: Ivan R. Dee, 2002), 2:551.

20. On the political uses of trees in Palestine, see Shaul Ephraim Cohen, *The Politics of Planting: Israeli-Palestinian Competition for Control of Land in the Jerusalem Periphery* (Chicago: University of Chicago Press, 1993), 44–50.

21. Alexandra Minna Stern, *Eugenic Nation: Faults and Frontiers of Better Breeding in Modern America* (Berkeley: University of California Press, 2005), 115–49.

22. Kathy J. Cooke, "Grant, Madison," American National Biography, http://www.anb.org.ezproxy.library.uvic.ca/articles/13/13-00642.html, accessed April 21, 2010; Matthew Press Guterl, *The Color of Race in America, 1900–1940* (Cambridge, MA: Harvard University Press, 2001).

23. Richard St. Barbe Baker, *My Life—My Trees* (London: Lutterworth Press, 1970), 75; Susan R. Schrepfer, *The Fight to Save the Redwoods: A History of Environmental Reform, 1917–1978* (Madison: University of Wisconsin Press, 1983).

24. Stern, *Eugenic Nation*, 124.

25. Richard St. Barbe Baker, *The Redwoods* (London: George Ronald, 1959 [1943]), 132.

26. Kevin Dann, *Across the Great Border Fault: The Naturalist Myth in America* (New Brunswick, NJ: Rutgers University Press, 2000), 130; William D. Stansfield, "Luther Burbank: Honorary Member of the American Breeders' Association," *Journal of Heredity* 97, no. 2 (2006): 95–99.

27. Baker, *The Redwoods*, 74–76; Baker, *My Life—My Trees*, 85.

28. "Proposed Clinic of Human Heredity," Eugenics Archive, http://www.eugenicsarchive.org/html/eugenics/static/images/1795.html, accessed April 21, 2010.

29. Baker, *My Life—My Trees*, 74.

30. Baker, *Men of the Trees: In the Mahogany Forests of Kenya and Nigeria*, xiv.

31. On similar organizations, see John Stevenson, "The Countryside, Planning and Civil Society in Britain, 1926–1947," in Jose Harris, *Civil Society in British History: Ideas, Identities, Institutions* (Oxford: Oxford University Press, 2003), 191–212.

32. See, for example, *Daily Mirror,* March 23, 1923, 10; July 6, 1929, 5; November 10, 1930, 18; November 3, 1933, 11; November 25, 1933, 4.

33. Baker, *My Life—My Trees.*

34. Introduction to Karen Gridley, ed., *Man of the Trees: Selected Writing of Richard St. Barbe Baker* (Willits, CA: Ecology Action, 1989).

35. Augustine Henry, *Forests, Woods and Trees in Relation to Hygiene* (London: Constable, 1919).

36. On materialist versus romantic portrayals of conservation in the United States, see Michael Williams, "The End of Modern History?," *Geographical Review* 88, no. 2 (1998): 275–300.

37. Richard St. Barbe Baker, "Tree and Shrub Life," in *Nature in Britain,* ed. Henry Williamson (London: B. T. Batsford, 1936), 156.

38. Richard St. Barbe Baker, *Among the Trees* (London: Privately printed by the Men of the Trees, 1935), 25, 35, 70.

39. Such terms were also bandied about by eugenically inspired novelists like Warwick Deeping, who has one character state, "Try to see that half the babies are not born. You don't let a garden get overcrowded with a lot of weedy rubbish. . . . The world wants cleaning and replanting,—but the drive of life is different in different plants. You have to allow for that. I would half the population, and try to see that the half that remained had a better chance." Warwick Deeping, *Sorrell and Son* (New York: Grosset and Dunlap, 1926), 251.

40. T. Earle Welby, *Figs from Thistles* (London: Philpott, 1925); Albert Edward Wiggam, *The Fruit of the Family Tree* (Indianapolis: Bobbs-Merrill, 1925).

41. See Lord Horder's contrast of "the eugenist" with the "the environmentalist" in "The Strain of Modern Civilization," *Nature*, September 26, 1936, 529–31.

42. On the "depredations of the Kikuyu," see D. E. Hutchins, *Report on the Forests of British East Africa* (London: HMSO, 1909); and for a defense of the natives' strategies, see Alfonso Peter Castro, "Indigenous Kikuyu Agroforestry: A Case Study of Kirinyaga, Kenya," *Human Ecology* 19, no. 1 (March, 1991): 1–18; James Fairhead and Melissa Leach, *Misreading the African Landscape: Society and Ecology in a Forest-Savanna Mosaic* (Cambridge: Cambridge University Press, 1996), 24–27.

43. Baker, *My Life—My Trees.*

44. Baker, *Men of the Trees: In the Mahogany Forests of Kenya and Nigeria,* 24. Timothy H. Parson, *Race, Resistance, and the Scout Movement in British Colonial Africa* (Athens: Ohio University Press, 2004).

45. Baker, *Men of the Trees: In the Mahogany Forests of Kenya and Nigeria*, 43.

46. Richard St. Barbe Baker, *The Brotherhood of the Trees* (London: Adelphi, 1930), 15, 34, 50. An anonymous review of *The Brotherhood of the Trees* noted that the idea of turning forest destroyers into conservationists was inspiring, but sounded too good to be true. *Geographical Journal* 77, no. 4 (April 1931): 390.

47. "A Short Account of the Men of the Trees or African Forest Scouts," reprinted from the *Empire Forestry Journal* 3, no. 1 (July 1924): 4, in Baker Papers, MG 71, Box 6; Baker, *Men of the Trees: In the Mahogany Forests of Kenya and Nigeria*, 84.

48. Philip Conford, *The Origins of the Organic Movement* (London: Floris Books, 2001), 61.

49. Ironically Cedric Dover, a Calcutta-born Eurasian, who in the 1930s wrote on eugenics, in the 1950s became editor of *Trees*. See Lucy Bland, "British Eugenics and 'Race Crossing': A Study of an Interwar Investigation," *New Formations* 60 (2007): 66–78.

50. Richard St. Barbe Baker, *Kamiti: A Forester's Dream* (London: George Ronald, 1958).

51. Baker, *The Redwoods*, 30.

52. Richard St. Barbe Baker, *Dance of the Trees (The Adventures of a Forester)* (London: Oldbourne Press, 1956), 76.

53. See also L. Cranworth, *Profit and Sport in Africa* (London: Macmillan, 1919), 78–79, 80, 85.

54. Baker, *My Life—My Trees*, 44. For the African response, see Parmenas G. Mockerie, *An African Speaks for His People* (London: Hogarth Press, 1934); and on the white settlers, see Christine S. Nicholls, *Elspeth Huxley: A Biography* (New York: Harper Collins 2002), 60; Chloe Campbell, *Race and Empire: Eugenics in Colonial Kenya* (Manchester: Manchester University Press, 2007).

55. Patrick Wright, *The Village That Died for England: The Strange Story of Tyneham* (London: Faber and Faber, 1995), 237; and on "alien, self-seeding conifers," see also C. E. M. Joad, *Diogenes; or, The Future of Leisure* (London: Kegan Paul, Trench, Trübner, 1928), 31.

56. Baker, *Among the Trees*, 90; and for similar warnings against planting the "wrong sorts of trees," see Clough Williams-Ellis, ed., *England and the Octopus* (London: Geoffrey Bles, 1928), 177.

57. "Inspiration from Trees," *Trees: Journal of the Men of the Trees* 1 (1936): 8; and on "native" vegetation being threatened by the "colonization" of the alien, see Arthur Tansley, *The British Islands and Their Vegetation* (Cambridge: Cambridge University Press, 1939), 1:243–45.

58. Banu Subramaniam, "The Aliens Have Landed! Reflections on the Rhetoric of Biological Invasions," in *Making Threats: Biofears and Environmental Anxieties*, ed. Betsy Hartmann, Banu Subramaniam, and Charles Zerner (Lanham, MD: Rowman and Littlefield, 2005), 135–48.

59. Jonah H. Paretti, "Nativism and Nature: Rethinking Biological Invasion," *Environmental Values* 7, no. 2 (1998): 183–92.

60. Catherine Brace, "Gardenesque Imagery in the Representation of Regional and National Identity: The Cotswold Garden of Stone," *Journal of Rural Studies* 15, no. 4 (October 1999): 365–76.

61. G. Groening and J. Wolschke-Bulmahn, "The Ideology of the Nature Garden: Nation-

alistic Trends in Garden Design in Germany during the early Twentieth Century," *Journal of Garden History* 12, no. 1 (1992): 73–80; Joachim Wolschke-Bulmahn, "The Search for 'Ecological Goodness' among Garden Historians," in *Perspectives on Garden Histories,* ed. Michel Conan (Washington, DC: Dumbarton Oaks, 1999).

62. On the history of this notion, see Stephen Daniels, "The Political Iconography of Woodland in Later Georgian England," in *The Iconography of Landscape,* ed. Denis Cosgrove and Stephen Daniels (Cambridge: Cambridge University Press, 1988), 43–82.

63. Baker, *Men of the Trees: In the Mahogany Forests of Kenya and Nigeria*, 273.

64. Baker, *Kamiti: A Forester's Dream.*

65. *Trees: Journal of the Men of the Trees* 2 (1938): 111; and see Grey Owl, *The Tree* (London: Lovat Dickson, 1936).

66. Donald B. Smith, *From the Land of Shadows: The Making of Grey Owl* (Saskatoon: Western Producer Prairie Books, 1990).

67. Such views were not uncommon. See Grant Rodwell, "Nature Enthusiasm, Social Planning and Eugenics in Australian State Schools, 1900–1920," *Journal of Educational Administration History* 29 (1997): 5.

68. Baker, *Among the Trees*, 8.

69. *The Times*, June 6, 1937, 13.

70. *Trees: Journal of the Men of the Trees* 3 (1939): 72.

71. Baker, "Tree and Shrub Life," 156; and on ambivalence toward technological change, see Bill Luckin, *Questions of Power: Electricity and Environment in Inter-War Britain* (Manchester: Manchester University Press, 1990).

72. Baker, *Men of the Trees: In the Mahogany Forests of Kenya and Nigeria*, 82.

73. Baker, *I Planted Trees*, 154.

74. Baker, *I Planted Trees*, 161.

75. *Trees: Journal of the Men of the Trees* 3 (1939): 144.

76. Baker, *Green Glory*, 70.

77. *Trees: Journal of the Men of the Trees* 3 (1939): 107, 108.

78. J. L. Finlay, "John Hargrave, the Green Shirts and Social Credit," *Journal of Contemporary History* 5 (1970): 56; David Prynn, "The Woodcraft Folk and the Labour Movement, 1925–1970," *Journal of Contemporary History* 8 (1983): 79–95; and on escape from the "unnatural existence" of industrial society, see Mark Drakeford, *Social Movements and Their Supporters: The Green Shirts in England* (New York: St. Martin's Press, 1997), 39.

79. John Hargraves, *The Great War Brings It Home: The Natural Reconstruction of an Unnatural Existence* (London: Constable, 1919), 323; and see also Joseph Francis Charles Craven, "Redskins in Epping Forest: John Hargraves, the Kibbo Kift Kin and the Woodcraft Experience" (PhD thesis, History Department, University College London, 1998).

80. John Hargraves, *The Confession of the Kibbo Kift* (London: Duckworth, 1927), 57, 79, 83, 284; and see also Leslie Allen Paul, *The Republic of Children: A Handbook for Teachers of Working-Class Children* (London: George Allen and Unwin, 1938).

81. Baker, *I Planted Trees,* 154.

82. Baker, *I Planted Trees*, 95; Baker, *Dance of the Trees*, 86.

83. MacGregor Knox, *Mussolini Unleashed, 1939–1944: Politics and Strategy in Fascist*

Italy's Last War (Cambridge: Cambridge University Press, 1986), 158. In fact, Mussolini's ruralist penchant was countered by his desire to industrialize Italy.

84. Baker, *My Life—My Trees*, 59.

85. Patrick French, *Younghusband: The Last Great Imperial Adventurer* (London: Harper Collins, 1994).

86. David Matless, "Nature, the Modern and the Mystic: Tales from Early Twentieth Century Geography," *Transactions of the Institute of British Geographers* 16 (1991): 272–86.

87. Younghusband, cited in Matless, "Nature, the Modern and the Mystic," 278.

88. Francis Younghusband, "The Sense of Society," *Sociological Review* 17 (1925): 1–13.

89. George Seaver, *Francis Younghusband: Explorer and Mystic* (London: John Murray, 1952), 350.

90. David Matless, "Aliens, Empire and Naked Climbing: An Eminent Geographer Remembered," *Journal of Historical Geography* 22, no. 1 (1996): 86–90.

91. Francis Younghusband, *The Heart of Nature; or, The Quest for Natural Beauty* (London: John Murray, 1921), 125.

92. Vaughan Cornish, *National Parks, and the Heritage of Scenery* (London: Sifton Praed, 1931); Vaughan Cornish, *Beauties of Scenery* (London: F. Muller, 1944).

93. Vaughan Cornish, *The Poetic Impression of Natural Scenery* (London: Sifton Praed, 1931), 9.

94. Vaughan Cornish, *The Preservation of Our Scenery* (Cambridge: Cambridge University Press, 1937), 18.

95. David Matless, "A Modern Stream: Water, Landscape, Modernism, and Geography," *Society and Space* 10 (1992): 569–88.

96. Vaughan Cornish, "The Geographical Aspects of Eugenics," *Eugenics Review* 16 (1924): 267–69; and see also K. Dodds, "Eugenics, Fantasies of Empire and Inverted Whiggism: The Political Geography of Vaughan Cornish," *Political Geography* 13 (1994): 85–99.

97. Baker Papers, MG 71, Box 4.

98. Suzanne Raitt, *Vita and Virginia: The Work and Friendship of V. Sackville-West and Virginia Woolf* (Oxford: Oxford University Press, 1993), 42.

99. Vita Sackville-West, *Heritage* (London: W. Collins Sons, 1919), 42–43, 243, 248.

100. Vita Sackville-West, *Dragon in Shallow Waters* (London: W. Collins and Sons, 1921).

101. Vita Sackville-West, *The Land* (London: W. Heinemann, 1927).

102. Philip Coupland, "The Blackshirt Utopians," *Journal of Contemporary History* 33, no. 2 (1998): 268.

103. Anne Williamson, *Henry Williamson: Tarka and the Last Romantic* (London: Alan Sutton, 1995), 222–25; A. W. Brian Simpson, *In the Highest Degree Odious: Detention without Trial in Wartime Britain* (Oxford: Clarendon, 1992), 215–16.

104. Baker, *I Planted Trees*, 225.

105. Baker, *Among the Trees*, 7; Baker, *I Planted Trees*, 78; "Our Society," *Trees: Journal of the Men of the Trees* 2 (1937): 1.

106. Malcolm Chase, "This Is No Claptrap, This Is Our Heritage," in *The Imagined Past: History and Nostalgia*, ed. Christopher Shaw and Malcolm Chase (Manchester: Manchester

University Press, 1989), 128–46; Peter Mandler, *The English National Character: The History of an Idea from Edmund Burke to Tony Blair* (New Haven, CT: Yale University Press, 2006), 148–55.

107. Stella Gibbons, *Cold Comfort Farm* (London: Penguin, 1944 [1932]), 95; and see also Faye Hammill, *Women, Celebrity, and Literary Culture between the Wars* (Austin: University of Texas Press, 2007), 152–78.

108. Evelyn Waugh, *Scoop: A Novel about Journalists* (London: Chapman and Hall, 1938), 26.

109. Evelyn Waugh spoofed the sort who joined the Society for the Protection of Rural England. "'Housing scheme,' 'Development,' 'Clearance,' 'Council houses,' 'Planning'—these obscene words had been expunged from the polite vocabulary of the district, only to be used now and then, with the licence allowed to anthropologists, of the fierce tribes beyond the parish boundary." Evelyn Waugh, "An Englishman's Home," *Tactical Exercise* (Boston: Little, Brown, 1939), 113.

110. Michael Bartholomew, *In Search of H. V. Morton* (London: Methuen, 2004).

111. Clough Williams-Ellis, ed., *Britain and the Beast* (London: Readers' Union, 1938).

112. Gerard Vernon Wallop, Viscount Lymington [Earl of Portsmouth], *Alternative to Death: The Relationship between Soil, Family and Community* (London: Faber and Faber 1943), 57, 67–73. J. R. R. Tolkien's work with its stress on the importance of trees represented a popular version of such views.

113. D. H. Lawrence, *Lady Chatterley's Lover* (New York: Grove Press, 1982 [1928]), 81. Strikingly enough Karel Čapek turned from writing about robots to noting how in England trees influenced Toryism by inspiring a respect for tradition. Karel Čapek, *Letters from England*, trans. Paul Sever (London: Geoffrey Bles, 1925), 15–16.

114. Lymington, *Alternative to Death*, 106.

115. Matless, *Landscape and Englishness*, 110, 115; Philip Conford, "Finance versus Farming: Rural Reconstruction and Economic Reform, 1894–1955," *Rural History* 13 (2002): 225–41.

116. Anna Bramwell, *Ecology in the Twentieth Century: A History* (New Haven, CT: Yale University Press, 1989), 125–30; R. J. Moore-Colyer, "A Voice Clamouring in the Wilderness: H. J. Massingham (1888–1952) and Rural England," *Rural History* 13 (2002): 199–224.

117. Richard Moore-Colyer and Philip Conford, "A 'Secret Society'? The Internal and External Relations of the Kinship of Husbandry, 1941–52," *Rural History* 15, no. 2 (2004): 189–296; R. J. Moore-Colyer, "Back to Basics: Rolf Gardiner, H. J. Massingham, and 'A Kinship in Husbandry,'" *Rural History* 12 (2001): 85–108.

118. Baker Papers, MG 71, Box 5.

119. H. J. Massingham, *The Small Farmer* (London: Collins, 1947), 63. And see also Jacquetta Hawkes, *A Land* (London: Cresset Press, 1951), 215.

120. R. G. Stapledon, *The Land: Now and Tomorrow* (London: Faber and Faber, 1935), 231.

121. R. G. Stapledon, *Disraeli and the New Age* (London: Faber and Faber, 1943), 149; and see also Pyrs Gruffudd, "Back to the Land: Historiography, Rurality and the Nation in Interwar Wales," *Transactions of the Institute of British Geographers* 19, no. 1 (1994): 64.

122. Albert Howard, *The Soil and Health: A Study of Organic Agriculture* (New York: Devin-Adair, 1947), 260.

123. Albert Howard, *The Maintenance of Soil Fertility* (London: Faber and Faber, 1945); Albert Howard, *An Agricultural Testament* (London: Oxford University Press, 1940); and see Michael Graham, *Soil and Sense* (London: Faber and Faber, 1941); Matthew Reed, "Fight the Future! How the Contemporary Campaigns of the UK Organic Movement Have Arisen from their Composting of the Past," *Sociologia Ruralis* 41, no. 1 (2001): 131–45; Ina Zweiniger-Bargielowska, "Raising a Nation of 'Good Animals': The New Health Society and Health Education Campaigns in Interwar Britain," *Social History of Medicine* 20 (2007): 73–89.

124. Howard, *An Agricultural Testament*, 18.

125. *Trees: Journal of The Men of the Trees* 11 (1947): 24.

126. J. B. S. Haldane, *Daedalus; or, Science and the Future* (London: Kegan Paul, Trench, Trübner, 1924), 39; and for an anarchist position on land reform, see George Woodcock, *New Life to the Land* (Bristol: Freedom Press, 1942).

127. Lord Northbourne, *Look to the Land* (London: Dent, 1940); David Smith and Malcolm Nicholson, "Nutrition, Education, Ignorance and Income: A Twentieth-Century Debate," *The Science and Culture of Nutrition*, ed. Andrew Cunningham and Harmke Kamminga (Amsterdam: Rodopi, 1995), 288–318.

128. Pierse Loftus, *The Creed of a Tory* (London: Philip Allen, 1926), 40–41.

129. Laurence Easterbrook, *British Agriculture* (London: British Life and Thought, 1943), 48.

130. G. M. Trevelyan, *Must England's Beauty Perish?* (London: Faber and Gwyer, 1929), 22; and see Peter Mandler, "Against 'Englishness': English Culture and the Limits to Rural Nostalgia, 1850–1940," *Transactions of the Royal Historical Society* 7 (1997): 155–75.

131. G. T. Wrench, *Reconstruction by Way of the Soil* (London: Faber and Faber, 1946), 13; and see also G. T. Wrench, *Wheel of Health* (London: C. W. Daniel, 1938); G. T. Wrench, *Restoration of Peasantries* (London: C. W. Daniel, 1939).

132. Richard Griffiths, *Fellow Travelers of the Right: British Enthusiasts for Nazi Germany, 1933–39* (London: Constable, 1980), 142–45; Georgina Boyes, *The Imagined Village: Culture, Ideology, and the English Folk Revival* (Manchester: Manchester University Press, 1993), 153–63; Conford, *The Origins of the Organic Movement*, 62.

133. Wright, *The Village That Died for England*, 211–40.

134. Dan Stone, "The English Mistery, the BUF, and the Dilemmas of British Fascism," *Journal of Modern History* 75, no. 2 (2003): 336; Moore-Colyer and Conford, "A 'Secret Society'?," 196–200. On D. H. Lawrence's interest in Gardiner, see David Bradshaw, "Red Trousers: Lady Chatterley's Lover and John Hargrave," *Essays in Criticism* 55 (2005): 352–73; *The Letters of D. H. Lawrence*, ed. Warren Roberts, James T. Boulton, and Elizabeth Mansfield (Cambridge: Cambridge University Press, 1987), 5:66, 293, 552–53, 591; 6:256–57.

135. Williams-Ellis, ed., *Britain and the Beast*, 284; and see Franz-Josef Brueggemeier, Marc Cioc, and Thomas Zeller, eds., *How Green Were the Nazis? Nature, Environment, and Nation in the Third Reich* (Athens: Ohio University Press, 2005); Thomas Zeller, *Driving*

Germany: The Landscape of the German Autobahn, 1930–1970, trans. Thomas Dunlap (New York: Berghahn Books, 2007).

136. Viscount Lymington [Gerard Vernon Wallop], *Should England Fight?* (London: British Council Against European Commitments, 1938); and see also Anna Bramwell, *Ecology in the Twentieth Century: A History* (New Haven, CT: Yale University Press, 1989), 113–18; Dan Stone, "The Far Right and the Back-to-the-Land Movement," in *The Culture of Fascism: Visions of the Far Right in Britain,* ed. Julie V. Gottlieb and Tom Linehan (London: I. B. Taurus, 2003), 182–99.

137. Lymington, *Alternative to Death,* 6, 21. T. S. Eliot hailed Lymington for pointing out the evils of the domination of "finance" and "the urbanizaton of mind." "A Commentary," *Criterion* (October 1938): 60.

138. For the same argument made by a member of both the Eugenics Society and the Men of the Trees, see Captain George Henry Pitt-Rivers, *Weeds in the Garden of Marriage* (London: Noel Douglas 1931), 74. Pitt-Rivers—a cousin of Churchill, a grandson of General Pitt-Rivers, and father of the Michael Pitt-Rivers famously arrested for a homosexual offense in 1954—went so far in expressing his pro-fascists views that he was interned during the war. For the judgment that he was "somewhere between eccentric and dotty," see Simpson, *In the Highest Degree Odious,* 217.

139. Gerard Vernon Wallop, Viscount Lymington [Earl of Portsmouth], *Famine in England* (London: H. F. and G. Witherby, 1938), 77, 87. Despairing of England ever seeing the light, Lymington retired to his farm in Kenya. Gerard Vernon Wallop, Viscount Lymington [Earl of Portsmouth], *A Knot of Roots: An Autobiography, by the Earl of Portsmouth* (London: Geoffrey Bles, 1965).

140. On the internment of Captain Thomas Guillaume St. Barbe Baker, MC, see Nicolas Hillman, "'Tell me chum, in case I got it wrong. What was it we were fighting during the war?' The Re-emergence of British Fascism, 1945–58," *Contemporary British History* 15, no. 4 (Winter 2001): 1–34; Simpson, *In the Highest Degree Odious*, 225.

141. Sheila Rowbotham, *Edward Carpenter: A Life of Liberty and Love* (London: Verso, 2009), 311–12, 444; and see also Raphael Samuel, *Island Stories: Unravelling Britain* (London: Verso, 1999), 132–52.

142. Dora Russell, *Tamarisk Tree 2: My School and the Years of War* (London: Virago, 1980), 78–79.

143. Jill Kathryn Benton, *Naomi Mitchison: A Century of Experiment in Life and Letters* (London; Boston: Pandora, 1990), 133–38.

144. On the postwar activists in the Soil Association, see Richard Moore-Colyer, "'Towards 'Mother Earth': Jorian Jenks, Organicism, the Right and the British Union of Fascists," *Journal of Contemporary History* 39 (2004): 353–71; and on the Greens' linkages to the romantic tradition, see Meredith Veldman, *Fantasy, the Bomb and the Greening of Britain* (Cambridge: Cambridge University Press, 1994).

145. Amir Weiner, ed., *Landscaping the Human Garden: Twentieth-Century Population Management in a Comparative Framework* (Stanford, CA: Stanford University Press, 2003).

146. Viscount Lymington, "The Policy of Husbandry," in *England and the Farmer: A Symposium,* ed. H. J. Massingham (London: Faber, 1941), 15.

147. Wright, *The Village That Died for England*, 206.

148. Stella Gibbons, *Cold Comfort Farm* (London: Penguin, 1944), 56, 59.

149. H. J. Massingham, *The Tree of Life* (London: Chapman and Hall, 1943), 207, 193.

150. Hugh S. Seaton, "How the Hills Breed Sound Stock," *Trees: Journal of the Men of the Trees* 10 (1946): 97–99.

151. Michael Graham, *Soil and Sense* (London: Faber and Faber,1941), 26.

152. On similar linkages elsewhere, see Roger Griffin, *Modernism and Fascism: The Sense of a Beginning under Mussolini and Hitler* (London: Palgrave, 2007), 6, 52.

CONCLUSION

1. See, for example, Anna Vaninskaya, "Janus-Faced Fictions: Socialism as Utopia and Dystopia in William Morris and George Orwell," *Utopian Studies* 14 (2003): 83–98.

2. George Orwell, *Coming Up for Air* (Harmondsworth: Penguin Books, 1962 [1939]).

3. George Orwell, *Nineteen Eighty-Four* (New York: Harcourt, Brace and World, 1949), part 2, chap. 2. Bernard Crick notes Eric Blair chose "George Orwell" as his nom de plume because it had "a manly, English, indeed country-sounding, ring to it." Bernard Crick, *George Orwell: A Life* (London: Martin Secker and Warburg, 1980), 147.

4. Andy Croft, "Worlds without End Foisted upon the Future—Some Antecedents of *Ninety Eighty-Four,*" in *Inside the Myth: Orwell: Views from the Left,* ed. Christopher Norris (London: Lawrence and Wishart, 1984), 183–216; Andy Croft, *Red Letter Days: British Fiction in the 1930s* (London: Lawrence and Wishart, 1990).

5. James Leslie Mitchell [Lewis Grassic Gibbon], *Gay Hunter* (London: William Heinemann, 1934).

6. Joseph O'Neill, *Land under England* (London: V. Gollancz, 1935).

7. John Palmer, *Hesperides: A Looking Glass Fugue* (London: Secker and Warburg, 1936), 61.

8. Naomi Mitchison, *We Have Been Warned* (London: Constable, 1935).

9. Storm Jameson, *In the Second Year* (New York: Macmillan, 1936), 73, 188.

10. In this bizarre story, the progressives are aided by tiny nonhuman seasonally sexed creatures who obey a mating list created by a Eugenics Society. Andrew Marvell [Howell Davis], *Minimum Man; or, Time to Be Gone* (London: Victor Gollancz, 1938).

11. Rex Warner, *The Aerodrome* (London: John Lane, 1941), 201, 310.

12. Katharine Burdekin, *Swastika Night* (London: Lawrence and Wishart, 1985 [1937]).

13. Katharine Burdekin, *Proud Man* (New York: Feminist Press, 1993 [1934]), 17, 25, 36.

14. Cyril Connolly, "Year Nine" (1938), in *The Condemned Playground: Essays: 1927–1944* (New York: Macmillan, 1946), 158.

15. Daphne Patai, *The Orwell Mystique: A Study in Male Ideology* (Amherst: University of Massachusetts Press, 1984).

16. George Woodcock, *Crystal Spirit: A Study of George Orwell* (London: J. Cape, 1966), 261, 284; Christopher Hitchens, *Why Orwell Matters* (New York: Basic, 2002), 141–54.

17. George Orwell, *Keep the Aspidistra Flying* (Harmondsworth: Penguin Books, 1968 [1936]), 151, 152.

18. George Orwell, *The Road to Wigan Pier* (New York: Harcourt Brace Jovanovich, 1972 [1937]), 248.

19. Sonia Orwell and Ian Angus, *The Collected Essays, Journalism, and Letters of George Orwell* (New York: Harcourt, Brace and World, 1968), 4:277–78. An equally gloomy view of Britain's demographic future was presented by Storm Jameson's husband, Guy Chapman, *Culture and Survival* (London: Jonathan Cape, 1940), 223–29.

20. Orwell and Angus, *The Collected Essays*, 3:48, 49.

21. William R. Steinhoff, *George Orwell and the Origins of 1984* (Ann Arbor: University of Michigan Press, 1975).

22. Orwell, *Nineteen Eighty-Four*, 65, 66, 270.

23. Cass R. Sunstein, "Sexual Freedom and Political Freedom," in *On "Nineteen Eighty-Four": Orwell and Our Future*, ed. Abbott Gleason, Jack Goldsmith, and Martha C. Nussbaum (Princeton, NJ: Princeton University Press, 2005), 233–41.

24. Hera Cook, *The Long Sexual Revolution: English Women, Sex, and Contraception, 1800–1975* (Oxford: Oxford University Press, 2004); Kate Fisher, *Birth Control, Sex, and Marriage in Britain, 1918–1960* (Oxford: Oxford University Press, 2006).

INDEX

Aarons, Jervois, 120
abortion: eugenic, 28; in future, 13; in dystopias, 159
Adam, Auguste Villiers de l'Isle, 69
Addison, Hugh, 172n96
adoption. *See* artificial insemination
Adler, Alfred, 41
Agate, James, 66
aging: eugenic breeding and, 89; hormonal treatments to counter, 84, 85, 86, 89, 96–97. *See also* hormones
agriculture, 149, 150, 151, 153. *See also* organic farming
airplanes, 38, and fertility control; 21–22 in future, 23, 37
Aldington, Richard, 59
Allen, Edgar, 101
Allenby, Edmund, 140, 143–44, 147
Allinson, Francesca, 122
America: conditioning in, 64, conservationists and environmentalists in, 140, eugenics in, 140
Americanization, threat of, 4, 139, 217n17
animals: experiments on, 87–88, 89, 102–3, 115, representations of, 17, 149–50
Arlen, Michael, 21, 50
Armstrong, Charles Wickstead, 74
Armstrong-Jones, Robert, 92
Arnold, Edwin Lester, 15
artificial insemination: adoption or, 128, 132; adultery and, 112–14,125, 129, 204n15; birth control and, 118, 131–32; defense of, 124–33; by donor, 117, 118, 124, 128–29, 132; eugenics and, 120, 127, 130–33, 215n179; fatherhood and, 119; feminists and, 121, 122, 128, 213n137; fertility and, 127, 131; fictional depictions of, 116, 121–22, 206n36, 215n178; in France, 115–17; in Germany, 117; history of, 30, 35, 109, 113–17; homosexuals use of, 122; hostility to, 115–16, 124–30; by husband, 123, 124, 129; in Italy, 117; legal status of, 2, 113–14, 124–25; lesbians and, 122, 133; marriages and, 127; masculinity

artificial insemination (*continued*)
and, 129–30; masturbation and, 116, 123, 126, 129; Medical Defense Union and, 125; medical profession and, 117–18, 119–20, 125–26; methods of, 115, 122–23, 205n29, 207n53; motherhood and, 119; and Orford vs. Orford trial, 110–14, 125, 204n17; religious views on, 116–17, 126; sex radicals and, 118; and sperm banks, 122, 131; success rate, 210n91; unmarried women and, 118–19, 124, 127–28 133, 213n133; in the United States, 115, 117, 122, 123, 131
Aschheim, Selmar: and the Aschheim-Zondek reaction, 100; and pregnancy tests, 101
Astor, Nancy (Viscountess Astor), 128
Atherton, Gertrude, 98
Auden, W. H., 60
Austin, F. Britten, 25
automatons. *See* robots
automobiles: accidents in, 41; adultery and, 43–44; advertising of, 41; Americans' access to, 42–43, 49; automobility and, 38; and birth control, 56, 183n127; chauffeurs and, 46–47; and class, 43, 44, 45–46; consumption and use of, 45, 46, 56; deaths caused by, 50–51, 55; fertility decline attributed to, 38, 54–55; freedom offered by, 42, 49; homosexuality and, 44–45; as icons of modernity, 37, 56; impotence and, 51–52; masculinity and, 38–42, 46; lesbians and, 47–48; potency and, 41; and seduction, 39–40, 41, 42, 43; and sexuality, 6, 176n4; sexual metaphors based on, 51, 52–53; speed of, 38–39; threats posed by, 39, 40, 45, 50; women and, 47, 48–51, 181n89; youths and, 42–43

Baker, Richard St. Barbe: American contacts, 139, 140; and Civilian Conservation Corps Camps, 141, 146; and colonial subjects, 142–43; and country life, 150–51; early career, 138–41; and empire, 143–44, 146–47; and Men of the Trees, 138, 141, 143, 147, 150, 153, 154, 156; nativist views, 144; portrayal of trees, 141–42, 145; publications of, 141, 149; and religion, 138, 148; response to 1930s Depression, 145–46; social views, 145, 150–51; supporters of, 147–54
Baker, Stanley L., 76
Baker, Thomas Guillaume St. Barbe, 154
Balfour, Evelyn, 147
Balint, Michael, 41
Banting, Frederick G., 87, 88, 89
Barton, Mary, 123, 124, 125, 128, 129, 131, 132
Bathurst, Benjamin (Viscount Bledisloe), 124
Beauvoir, Simone de, 180n79, 213n137
Bell, George (Bishop of Chichester), 124
Bellamy, Edward, 11, 26
Benjamin, Harry, 94, 98
Berman, Louis, 93, 104, 106–7
Bernal, John Desmond, 3, 30, 32, 72, 109
Berthold, Arnold Adolf, 83, 94
Besant, Walter, 12
Best, Charles H., 87
Bibby, Cyril, 131
Bird, Alfred, 44
Birnstingl, Harry Joseph, 24, 50
birth control: impact of, 21, 70; popularization of, 15, 18, 21; and the Society for Constructive Birth Control and Racial Progress, 18; as symbol of modernity, 2, 21, 29, 155
Blacker, C. P., 3, 19, 20; on artificial insemination, 128, 131; on class, 35; on fertility, 27; on sterilization, 71
blackmail, 44–45, 125
Bogdanov, Alexander, 171n80
Booth, Florence, 43
Bowen, Elizabeth, 51
Boy Scouts, 142, 146
Brave New World, 5, 9, 10; critics' response to, 16–17, 36; family in, 24; fertility control in, 20; men in, 15; women in, 14. *See also* ectogenesis; Huxley, Aldous
Brewer, Herbert, 131
Brittain, Vera, 3; on ectogenesis, 35, 109; on future, 22–23; on hormones, 98
Britten, Florence, 42
Bromley, Dorothy, 42
Broster, Lennox Ross, 103–4, 106
Brown-Séquard, Charles-Edward, 83, 90
Browne, Stella, 122

Browne, Thomas, 114
Buchan, John, 49
Bulwer-Lytton, Edward George, 13
Burbank, Luther, 140
Burdekin, Katharine, 159
Butenandt, Adolf, 89, 90, 101
Butler, Samuel, 188n66

Cadbury, Elizabeth, 147
Campanella, Tommaso, 5
Campbell, M. L., 68
Čapek, Karel, 4, 57, 62, 68, 73, 74, on trees, 222n113. *See also* robots
Carpenter, Edward, 154
Carrel, Alexis, 31, 33, 88
Carstairs, Marion "Joe," 180n80
Carter, C. O., 131
Catholic Church, artificial insemination and, 116–17, 119, 124, 126
Cavendish-Bentinck, William, 60
Cecil, Henry, 61
Chamberlain, Neville, 93
Chapman, Guy, 226n19
Charles, Edward (Charles Edward Hempstead), 176n166
Charles, Enid, 3, 21, 29, 35, 54
chauffeurs, 46–47
Chesser, Elizabeth Sloan, 70, 72
Chesser, Eustace, 127
Chesterton, G. K., 59
children: in dystopias, 25; needs of, 21; raising of, 24; resulting from artificial insemination, 119, 120, 125, 127; in state nurseries, 24, 25
Church of England, artificial insemination and, 126; birth control and, 55
Clarke, Frederick le Gros, 28
Clarkson, W., 190n93
Clouston, Joseph Storer, 69–70
Coats, Thomas (Lord Glentanar), 124
Comte, Auguste, 116
conditioning, 9, 17; and behavioral psychology, 24, 64
Connolly, Cyril, 159–60
conservation: eugenics and, 140, 142, landowners and, 150–53; leftists and, 154. *See also* Baker, Richard St. Barbe; trees
Cooper, Diana, 120
Corelli, Marie, 97–98
Cornish, Vaughan, 148
Coward, Noel, 82–83, 88–89, 182n105
Cox, Gladys, 52
craftsmanship, 154
Crew, F. A. E., 82, 94, 100–101, 102, 107, 193n2
Crichton-Miller, H., 63
criminals, 74, 192n136; hormonal treatment of, 91–92
cross-dressing, 13, 47

d'Albe, E. E. Fournier, 32
Darwin, Charles, 10
Darwin, Charles Galton, 55, 94
Darwin, Leonard, 71, 73, 76
Day-Lewis, C., 60
Deeping, Warwick, 218n39
degeneration, 27, 28, 39, 65
Dickinson, Robert Latou, 117
divorce, 24, 43–44, 112, 211n100
Dixie, Florence, 13
Dobbie, Beatrice, 128
Döderlein, Albert, 117
Doisy, Edward A., 101
Douglas-Hamilton, Nina (Duchess of Hamilton), 88
Dover, Cedric, 219n49
Doyle, Arthur Conan, 88
Dreyer, N. B., 107
drugs: in future dystopias, 15, 16
Drury, Newton B., 140
Duff, Ursula Grant, 1, 138, 142, 147
Duhamel, Georges, 49, 122, 177n8
Duncan, Isidora, 49
Dunsany, Lord (Edward Plunkett), 88
Durant, Henry, 55–56

Earengey, William George, 127, 131, 212n129
Easterbrook, Laurence, 153
ectogenesis: babies in bottles as, 35; critiques of, 33–34; in future society, 23, 31–34, 109; incubators and, 33; interest in, 30–35; race improvement by, 30
Eddy, James G., 140
Eliot, T. S., 39–40, 60, 101, 139, 224n137
Elkan, Rudolph, 200n144

Ellis, Havelock, 74, 118
endocrinology. *See* hormones
English Mistery, 153
eugenics: artificial insemination and, 120, 127, 130–33; birth control and, 19; and conservation movement, 140, 142; critiques of, 29–30, 75–76, 191n112; euthanasia and, 27; fertility differential and, 27, 66–67; futurist fantasies of , 25, 158; gender and, 68; hormones and, 84, 93, 105–7, 108; history of, 3–4, 5; immigration and, 144; negative policies of, 27, 71–75; positive policies of, 26–27, 70; progressives and, 29; selective breeding and, 70–72; and sterilization, 28, 71; working class and, 66, 154
Eugenics Society, 102, 131, 132, 138
euthanasia, 21, 27, 28, 73; British Voluntary Euthanasia Society and, 74
Ewers, Hans Heinz, 200n146

Fairfield, Letitia, 124–25, 126, 127, 129
family. *See* futurist fictions
fatherhood, artificial insemination and, 119; future of, 24
feminism, and eugenics, 34–35, 67, 70–71, 75–76; and futurist literature, 159, 167n19
Ferguson, Adam, 187n44
fertility, class differences in, 66–67; decline of, 21
films, portrayal of future in, 122, 169n43
Foote, Edward Bliss, 211n108
Forbes, Robert, 131
Ford, Henry, 37, 63, 64; Fordism and, 63, 64
Forster, E. M., 39, 45
Fox-Davies, A. C., 12
France, 115–17
Frankenstein's monster, 30, 62, 63, 66
Frankau, Gilbert, 41
Freeman, R. Austin, 25, 39, 64–65, 71
freemartins, 81
Freud, Sigmund, 41, 51, 86–87
Fuller, J. F. C., 59
Fülop-Miller, René, 6
futurist fictions: artificial insemination in, 161; children in, 18, 24, 25, 159; ectogenesis in, 23; and family forms, 24; fascism in, 158–59; and feminist views, 13, 15; fertility control and, 18; homosexuality in, 16; marriages in, 12, 26; optimistic, 10–11; progressives and, 158–59; reproduction in, 12, 19, 26, 31; sexually aggressive women in, 13; sexual relations in, 5; socialism depicted in, 11–12, 25; wars in, 14; women's position in, 11, 12, 14, 25. *See also* science fiction

Gallachan, Walter M., 54
Galsworthy, John, 54
Galton, Francis, 25–26, 149
Gardiner, Rolf, 153
Gastev, Alexey, 64
Gates, Reginald Ruggles, 1, 70, 138, 142
Gautier, Jules, 116
Gayton, Bertram, 87
General Medical Council, 125
Germany, 4–5, 29, 144, 153, 154
Gérard, J., 115
Gibbon, Lewis Grassic, 14
Gibbons, Stella, 47, 150, 155
Gibbs, Philip, 92, 94, 139
Girault, Louis, 115
glands, endocrine. *See* hormones
Gloag, John, 17, 21, 50, 96–97, 171n74
Godwin, George, 27, 28, 92
Gordon-Watson, C., 125
Graham, Michael, 156
Grahame, Kenneth, 40
Grant, Madison, 140
Greene, Graham, 43, 44, 46
Gresswell, Elise Kay, 13
Grey Owl (Archibald Belaney), 145
Griffin, Bernard (Archbishop of Westminster), 126
Griffith, Edward Fyfe, 123

Hailie Selassie, 93
Haire, Norman: on artificial insemination, 120; on ectogenesis, 32, 109; on
homosexuality, 105–6, 122; on rejuvenation therapies, 4, 86, 87, 94; on sex change operations, 103; on sexual relations, 24, 100; on sterilization, 28

Haldane, Charlotte, 3, 16, 20, 21, 29; on ectogenesis, 109; on hormones, 89; and race, 34, 71; on sexual ignorance, 52–53
Haldane, J. B. S.: on aging, 97; on ectogenesis, 4, 31, 32, 35, 109; on eugenics, 71; on factory workers, 152; on hormones, 92; on sterilization, 29,
Hall, A. Daniel, 61
Hall, Radclyffe, 47–48
Hamilton, Cicely, 14
Hamilton, Patrick, 41, 178n27
Harbou, Thea von: *Metropolis*, 63, 69
Hargraves, John, 146
Hatfield, H. Stafford, 63
Heape, Walter, 30, 117
Heard, Gerald, 91
Heine, Heinrich, 34
Helmholtz, Hermann von, 62
Henry, Augustine, 141
Herbert, A. P., 54
Heron-Allen, Edward, 102–3, 201n147
Hill, W. P. H., 151–52
Hilton, James, 17–18
Hirschfeld, Magnus, 69, 105, 1117
Hitler, Adolf, 150, 158
Hodgkinson, George Edmund, 112–13, 203n9
Hoffman, E. T. A., 61–62, 68
Hogben, Lancelot, 72, 101, 191n112
Holtby, Winifred, 56
home: and the automobile, 50; middle-class, 20, 23, 24; and the socialist barrack, 12, 155
homosexuality: artificial insemination and, 122; automobiles and, 44–45; as "cancer," 148; causes of, 201n161; in futurist literature, 16, 158, 159; hormonal treatment of, 104–5. *See also* lesbians
Horder, Thomas, 45
hormones: and animal grafts, 84, 87–88; class and, 93–94; commercialization of, 90–91, 100, 101; and criminality, 91–92; discovery of, 81, 83, 87; eugenics and, 93, 105–7, 108; female, 83, 97–101; and fertility control, 102; fictional portrayals of, 82–83, 87–89, 94, 97–98, 99; in future societies, 23, 81; gender and, 94, 95; homosexuality and, 104–6; male, 83, 89–90, 95–96; mental defectiveness and, 92–93, 197n73; and pregnancy tests, 100–101; public's interest in, 82; race and, 93; replacement therapies, 2, 102; to restore potency, 6, 86; to retard aging, 84, 85, 86, 89, 96–97; and sex change operations, 103–4; and sexuality, 99, 102; and social control, 82, 91, 101. *See also* Steinach, Eugen; Voronoff, Serge
Housman, Laurence, 12
Howard, Albert, 152
Hunter, John, 115
Hutton, Isabel, 53
Huxley, Aldous: and America, 139–40; and automatons, 64; and automobiles, 37, 38, 41; *Brave New World*, 5, 9, 10, 14, 19–20, 36, 137; critics of, 36, 89, 137; and ectogenesis, 9, 30, 31–32, 109; and eugenics, 72–73, 76; and euthanasia, 73; and fertility control, 20, 67, 170n65; and future society, 16–17; on gender relations, 14, 168n32; on hormones, 81, 85, 94, 97; and sex, 3; and social conditioning, 37, 137; and workers, 176n163; works of, 9–10
Huxley, Julian, 3, 27; on artificial insemination, 131–32; on endocrinology, 33, 83, 107; on fertility differential, 30; on homosexuality, 104
Huxley, Thomas, 10
Huysmans, J. K., 69

impotence, 85, 91, 93, 95
industrial psychology, 62, 63
Inge, William, 66–67
Isaacs, Susan, 72
Italy, 117, 147, 153

Jackson, Margaret Hadley, 123, 127, 128, 129, 130, 132
Jaeger, Muriel, 21
James, Norah C., 50–51
Jameson, Storm, 158
Jeans, James, 71–72
Jenkins, Alice, 55
Jenks, Jorian, 147
Jews, anti-Semitism and, 149, 151, 153, 154, 158, 216n180

Joad, C. E. M., 3, 15; on America, 139; on automobiles, 39, 45; on machines, 65; on hormones, 91
Johnson, Owen, 33
Johnstone, R. W., 100

Kammerer, Paul, 94
Keane, C. F., 126
Keith, Arthur, 84, 93
Keller, David H., 98
Kendall, John (Margaret Maud Brash), 172n97
Kenya, 143–44; and Mau Mau, 145
Kibbo Kift Kin, 146
Kinship in Husbandry, 151, 152

Lacoste, René, 59
Lajartre, Oscar de, 115–16
Lane, C. Rickword, 130
Lang, Fritz, 69
Lankester, E. Ray, 102
Laqueur, Ernst, 89–90
Laski, Frida, 72
Lawrence, D. H.: on automobiles, 41–42, 55; on lethal chamber, 73; on reproduction, 121–22; on robots, 60; on trees, 151; on women, 42, 168n32
Lawrence, T. E., 140
Leacock, Stephen, 12, 68
Leavis, Q. D., 36
Lee, Vernon (Violet Paget), 28–29
Lempicka, Tamara de, 48–49
Lenin (Vladimir Ilyich Ulyanov), 64
lesbians: automobiles and, 47–48, 180n80; in futurist fiction, 159; increase in number, 99
Lesley, Shane, 147
lethal chamber, 26, 73
Levitt, Dorothy, 181n85
Lewis, Wyndham, 60
Lichtenstern, Robert, 104, 105
Lind-af-Hageby, Louise, 88
Loeb, Jacques, 30–31, 102
Loftus, Pierse, 153
Long, Norman, 54
Louis, Tracy, 47
Low, A. M., 64
Ludovici, Anthony, 3, 33, 39, 71, 76; on artificial insemination, 122, 128
Lydston, G. Frank, 214n155

Mace, David R., 52, 53
Macfie, Ronald Campbell, 63, 169n44
Maeterlinck, Maurice, 84
Mairet, Philip, 74, 147
Malcolm, John, 76
Malleson, Joan, 128, 129, 130, 213n139
Mannin, Ethel, 87
Manning, Laurence, 35
Mantegazza, Paolo, 117
Marañón, Gregorio, 104
March, Nora, 70
marriage: artificial insemination and, 116, 124, 126, 127, 130; dissolution of, 26; in future, 11, 15, 21, 22, 23, 28; group, 22; by lottery, 15; and marriage contracts, 13; "semi-detached," 23
Marshall, Alfred, 62
Marvell, Andrew, 158
Massingham, Hugh J., 144, 151, 152, 155
masturbation, 40, 83, 116. *See also* artificial insemination
McCandless, Anne Ethel, 127
Medico-Legal Society, 127, 131
Men of the Trees. *See* Baker, Richard St. Barbe
Metchnikoff, Elie, 96
Millard, C. Killick, 74
Mills, Elliot Evans, 12
Mitchell, James Leslie, 158
Mitchell, M. E., 99
Mitchison, Naomi (née Haldane), 3, 154, 158
Mjøen, Jon Alfred Hansen, 93
modernization: class differences and, 165n4; and modernity, 156; progressives and, 157; responses to, 2, 3, 10
Moll, Albert, 101
Money, Leo Chiozza, 59
Moore, John C., 40
Moore-Brabazon, John (Lord Brabazon), 124
Morris, William, 11
Morton, H. V., 151, 153
Mosley, Oswald, 150
motorcars. *See* automobiles
Mühsam, Richard, 105

Muller, H. J., 32, 109, 13, 174n133, 175n151
Mullins, Claud, 52, 72
Muscio, Bernard, 63, 187n46
music halls, 43, 45–46, 179n63
Mussolini, Benito, 93, 147, 148

names, replaced by numbers, 25, 29
National Research Foundation for the Eugenic Alleviation of Sterility. *See* Seymour, Frances
Nazis, 29, 72, 130, 138, 153, 159
Needham, Joseph, 19, 35
Neville-Rolfe, Sybil, 70, 72, 92–93, 105
Newman, Robert (Lord Mamhead), 147
Newsholme, H. P., 126
Newsom, G. E., 24
Newte, Horace Wykeham Can, 24–25
Nicolson, Harold, 149
Nordau, Max, 16
nudism, 19

O'Brien, Edward J., 60, 67
Odle, E. V., 57, 69
O'Neill, Joseph, 158
Orde, John F., 110, 112, 114
Orford, Frederick, 110–12, 114, 129
Orford, Lillian, 110–14, 118, 123, 129
organic farming, 151, 152, 153
organotherapy, 83, 87
Orwell, George (Eric Blair): consumerist society and, 56, 157; and *Nineteen Eighty-Four,* 130, 161; population concerns of, 160; reproduction and, 130, 160–61; and rural life, 157–58; on totalitarianism, 157

Palestine, 140, 143–44, 146
Palmer, John, 158
Pancoast, William, 207n46
parthenogenesis, 30, 35
Paul, Cedar, 86
Paul, Eden, 23–24, 86
Pearson, Karl, 25
Philips, J. G. Porter, 93
Piddington, Marion Louisa, 119
Pitt-Rivers, George Henry, 147, 154, 224n138
Plato, 26
Plunkett, Horace, 147
Ponsonby, Frederick Edward Neuflize (Earl of Bessborough), 147
Popenoe, Paul, 183n120
Pound, Ezra, 101, 107
Priestly, J. B., 65
prosthetics, 62, 82
prostitutes, 14, 69, 103, 179n46; "amateur," 20; male, 24
Proust, Marcel, 180n69

race, 18; hormones and, 93; protection of, 148–49. *See also* eugenics
Rachilde (Marguerite Eymery), 68–69, 189n85
Ragg, Lonsdale, 145
Rattenbury, Alma and Francis, 47
Rayner, W., 113
Read, Carveth, 74
Reeth, Allan, 12
Rejuvenation. *See* hormones
Richardson, Benjamin Ward, 73
Robinson, William J., 51–52, 106
robots: art and, 60; automatons and, 61–62, 63, 69; craftsman replaced by, 59, 62, 65; critics of, 59; domestic servants and, 37, 64; as epitome of machine culture, 59, 62; eugenics and, 6, 66, 67–68, 70–73; euthanasia and, 73–74; factory workers as, 35, 63, 64; fertility decline and, 58, 66, 67; in films, 63, 189n91; gender of, 68–69; as icons of modernity, 57, 76–77, 184n3; as portrayed in *R.U.R.*, 58, 62, 65, 66, 68, 74; reproduction of, 58, 65, 66, 109, 188n66; revolt of, 58, 64, 65, 186n38; sex and, 68–70, 75–76, 190n94; on stage, 60; as unthinking drones, 59–60, 61, 63, 187n43
Rockefeller Foundation, 102
Rolleston, Humphrey, 76
Roosevelt, Franklin Delano, 146
Royden, Maud, 147
Ruck, Berta, 47
Ruskin, John, 141
Russell, Bertrand: on automobiles, 53; on ectogenesis, 32–33; on eugenic breeding, 70, 72; on family forms, 24; on fertility differential, 29–30; on hormones, 91, 92; life of, vii, 3, 101; on sex, 17; suspicion of experts, 28, 32–33, 91, 92; on youth, 43

Russell, Christobel, 82
Russell, Dora, 3, 15, 29, 154, 168n35
Russell, John, 82
Ruzicka, Leopold, 90

Sackville-West, Vita, 149
Saklatvala, Shapurji, 65
Saleeby, C. W., 147
Samuel, H. L., 12–13
Sanger, Margaret, 53
Sayers, Dorothy L., 92
Schäfer, Edward. *See* Sharpey-Schäfer, E. A.
Scharlieb, Mary, 98–99
Schiller, F. C. S., 28
Schmidt, Peter, 86
science fiction, 2, 5, 65, 162; national differences in, 166n5
scientific management. *See* Taylor, Frederick, and Taylorism or scientific management
sex: "automobilic," 38, 51; female desire and, 96, 99; hormones and, 94; as means of social control, 17; sex change operations and, 103–4; and sex determination, 32, 91, 203n2; and sex education, 24; and sexual prudery, 17; state's interest in, 159, 161
Seymour, Frances, 131
Shanks, Edward, 14
Sharpey-Schäfer, E. A., 192n140
Shaw, George Bernard, 46, 59, 66, 70, 74, 84
Shelley, Mary, 30, 62, 66
Shelley, Percy Bysshe, 61
Sims, James Marion, 115
Sitwell, Osbert, 38–39, 42, 96
Smith, Frederick Edwin (Lord Birkenhead), 27, 28, 34, 97
Snow, C. P., 94
Soviet Union, 4, 15, 21, 64, 73, 88, 172n91
Spallanzani, Lazzaro, 115, 186n35
sperm, and sperm donors, 109, 124, 128–29
Stabb, Arthur F., 113
Stapledon, George, 152
Stapledon, Olaf, 16, 21–22, 28, 35
Starling, E. H., 84, 89
Steinach, Eugen: career, 85, 87, 97; experiments, 95; on femininity, 99–100; and homosexuality, 104, 105; opposition to, 89, 94; on rejuvenation, 86, 96; on sex change operations, 103
"Steinach Operation," 85; Benjamin on, 94; Freud on, 86–7; and impotence, 82, Walker on, 86, Yeats on, 86–7
sterilization: enforced, 26; of the feeble-minded, 28, 29, 30, 71–72, 138; in future, 23; methods of, 28; voluntary, 27; and Workers Committee to Legalize Voluntary Sterilization, 72
Stoker, Bram, 177n24
Stopes, Marie: artificial insemination and, 118–20, 127, 128; birth control and, 4, 18, 54, 156; on fertility differential, 67; on impotence, 52; marriage of, 1
Strahan, S. A. K., 71
streamlining, 66
Sturt, George, 45
Summerskill, Edith, 128, 213n141
Sutherland, Halliday G., 119
Swanwick, Helena, 67, 75–76
Sweven, Godfrey (John Macmillan Brown), 11–12, 13

Taylor, Frederick, and Taylorism or scientific management, 10, 62, 64
Tennant, Stephen, 208n72
test-tube babies, 6, 109. *See also* artificial insemination; ectogenesis
Thomas, Lowell, 140
Thouret, M. A., 205n26
thyroid, therapeutic aspects of, 84, 89
Tiller girls, 89, 185n21
Tippett, Michael, 122
"To-day and To-morrow Series," 63, 169n37
Tolkien, J. R. R., 60, 222n112
Tonks, Henry, 60
trains, 37, 38
transplantation: of organs, 53, 82, 99; of testicles, 83, 86
transport. *See* airplanes; automobiles; trains
trees: California redwoods, 140; and class,145–46; conservation of, 137; and deforestation,139, 142, 150, 151; eugenics and, 138, 142; and forestry, 139; native and alien, 144; romantic portrayals of, 141–42, 148; Royal Forestry Society and,

141; spiritual aspects of, 146, 148. *See also* Baker, Richard St. Barbe
Trevelyan, G. M., 153
Turner, H. M. Stanley, 129

United States of America. *See* America
urban life: impact on population, 152, 153, 154–55, 155–56; threat of, 145, 148, 149

Vacher de Lapouge, Georges, 116, 206n41
Vecki, Victor, 52
Velde, T.H. van de, 122–23
Victoria, Vesta, 43
Voronoff, Serge: and "monkey gland operations," 84, 87; opposition to, 88, 89, 107; on rejuvenation, 91, 96, 97

Walker, Frank, 93
Walker, Kenneth: and artificial insemination, 124, 126; on homosexuality, 106; on rejuvenation, 86, 96; on sex education, 53
Wallop, Gerard Vernon (Viscount Lymington), 153–54, 155, 224n139
Walshe, F. M. R., 130
Walton, Anne Elizabeth, 100
Walton, Arthur, 107
Warner, Rex, 158–59
Waterhouse, R. Denne, 40
Watson, John B., 24
Waugh, Evelyn, 40, 43, 44, 46, 48, 49; on countryside, 150, 222n109
Wegener, Einar (Lili Elbe), 103
Wellock, Wilfred, 56, 65
Wells, H. G., 3, 10; and automobiles, 45, 47; and cautionary tales, 33; and euthanasia, 73–74; and fertility, 18–19, 20–21; and future family, 11, 16, 22, 70; and hormones, 94; and misfits, 26, 27; racism of, 45; on sex, 17, 171n84; and transport, 37; views on women, 14, 15, 26, 168n28
West, Rebecca, 124
Weston, Mary (Mark Weston), 103–4
Whetham, W. C. D. and Catherine D., 67
White, Peter, 112
Wiesner, Bertold Paul: artificial insemination and, 210n92, 213n142; on homosexuality, 106; pregnancy tests and, 100–101; research of, 101–2, 200n140
Wilde, Oscar, 104
Williamson, Henry, 149–50, 154
Wilson, Alfred, 85–86
Wilson, Jesse, 12
Wodehouse, P. G.,
Woodcock, George, 36
Woodcraft Folk, 146
Woolf, Vanessa, 48
Woolf, Virginia, 48, 49–50, 121, 149
women: biological inferiority of, 95–96; duty to reproduce, 67, 70; emancipation of, 2, 23–24; future political power of, 13, 98; hormonal treatment of, 101, 102; and motherhood, 20; position of in utopias, 11, 12, 14; rejuvenation of, 95, 98–100
World War One, 95, 96, 108
World War Two, 161
Wrench, G. T., 153
Wright, Sydney Fowler, 21, 29, 55, 67

Yeats, William Butler, 86–87
Younghusband, Francis, 147–48, 182n113
Young, Meredith, 119

Zamiatin, Evgenii Ivanovich, 167n12
Zondek, Bernhard. *See* Aschheim, Selmar